# Handbook of Environmental Contaminants:

## A Guide for Site Assessment

# Handbook of Environmental Contaminants

## A Guide for Site Assessment

## Chris L. Shineldecker

**CRC Press**
Taylor & Francis Group
Boca Raton London New York

CRC Press is an imprint of the
Taylor & Francis Group, an **informa** business

First published 1992 by Lewis Publishers, Inc.

Published 2019 by CRC Press
Taylor & Francis Group
6000 Broken Sound Parkway NW, Suite 300
Boca Raton, FL 33487-2742

© 1992 by Taylor & Francis Group, LLC
CRC Press is an imprint of Taylor & Francis Group, an Informa business

First issued in paperback 2019

No claim to original U.S. Government works

ISBN-13: 978-0-367-45036-6 (pbk)
ISBN-13: 978-0-87371-732-8 (hbk)

Visit the Taylor & Francis Web site at
http://www.taylorandfrancis.com

and the CRC Press Web site at
http://www.crcpress.com

Library of Congress Card Number 91-44070

## Library of Congress Cataloging-in-Publication Data

Shineldecker, Chris L
    Handbook of environmental contaminants: A guide for
site assessment / Chris L Shineldecker
      p  cm
    Includes bibliographical references and index
    ISBN 0-87371-732-5
    1 Pollutants—Analysis 2 Hazardous wastes—Environmental
aspects 3 Factory and trade waste—Environmental aspects
1 Title

TD193 S55 1992
628 5'2—dc20
                                        91-44070
                                        CIP

TO MY WIFE JANICE,
FOR SHARING HER DREAMS WITH ME

***Chris L. Shineldecker*** is a Certified Professional Geologist with experience in the environmental assessment and investigation fields. He graduated from Central Michigan University with a B.S. degree in geology in 1980. Since then, Mr. Shineldecker has been involved in domestic and international assessments for real estate transactions and underground storage tanks, Superfund cleanups, hydrogeologic investigations, and petroleum exploration and production sites. Mr. Shineldecker is employed by the Environmental Response Division of the Michigan Department of Natural Resources. Mr. Shineldecker serves on the Environmental Site Assessment Committee of the Michigan Association of Environmental Professionals.

# PREFACE

Recent changes in environmental law have significantly impacted purchasers and owners of environmentally impaired real estate. These laws have also impacted lending and real estate institutions. The financial liability created by these laws has caused an explosion in the environmental assessment market as potential real estate purchasers and lenders attempt to protect themselves through the exercise of "all-appropriate-inquiry" and "due diligence." Large numbers of papers and books are being published discussing the need for and mechanics of performing environmental site assessments. Seminar attendance has skyrocketed. The frequency and duration of discussions that are taking place in board rooms and legal offices across the country all indicate that the magnitude of the costs of environmental liability is a sobering and formidable factor in today's business environment.

The Phase I environmental assessment has become the most common vehicle used to address the all-appropriate-inquiry and due diligence issues. Many excellent references are available that discuss how to carry out Phase I assessment and how to identify signs of potential environmental liability. Standards for the performance of these initial assessments are being addressed at the local, state, and federal levels. Much is known of the characteristics of thousands of chemicals that are potential environmental contaminants. Little work has been done, however, to compile a comprehensive reference source of chemicals that should or could be reasonably associated with specific industries, activities, manufacturing processes, and products.

This book is designed to serve as a such a reference to anyone who is charged with the task of performing, evaluating, and/or counseling regarding the performance of all-appropriate-inquiry and due diligence as they relate to environmental liability. The information presented in this reference is a compilation of selected industrial and manufacturing data, data compiled from various chemical reference works, published articles, occupational health and safety studies, and information obtained through various public agencies.

It would be impossible to produce a reference book that is 100% applicable for every facility or property. Use of this reference book, in conjunction with historical records, chemical reference compilations, interviews with knowledgeable people, and information about current and historical site activities will significantly facilitate the determination of potential environmental contaminants which may be present at a property and which could represent significant financial liability to the owner or future owners of that property.

CHRIS L. SHINELDECKER, CPG

# TABLE OF CONTENTS

# INTRODUCTION

The environmental movement first arose in the 1970s attempting to address the gross pollution of our water and air resources. Environmental laws and regulations passed by the federal government and various states have evolved in a manner that significantly impacts both the public and private sectors. The Comprehensive Environmental Response Compensation and Liability Act (CERCLA) was passed into federal law by Congress in 1980 to provide a regulatory structure to address the uncontrolled release of hazardous materials to the environment, particularly those caused by past activities. The Superfund Amendments and Reauthorization Act (SARA) was passed into federal law in 1986 to provide additional funding for government-sponsored cleanups. Other significant federal laws include the Resource Conservation Recovery Act (RCRA), the Clean Water Act (CWA), and the Clean Air Act (CAA).

Many states have adopted legislation based on the federal environmental statutes. These state and federal statutes impose liability not only for improper discharges to land, air, surface and ground water resources, and public sewer systems, but also for the improper handling, storage, and disposal of hazardous wastes.

Liability under various environmental statutes is broadly incurred by a number of potentially responsible parties (PRPs). PRPs include the current and/or past owners or operators of a property, any person who arranged for the treatment or disposal or arranged with a transporter for the transport for disposal or treatments of hazardous materials, and any person who accepts or accepted any hazardous materials for transport to disposal or treatment facilities. Responsible parties have been held liable for cleanup costs, damages for injury or loss of natural resources and assessment costs, and the costs of required health risk assessments.

The scope of liability for environmental problems has significant implications in the area of real estate transactions. Current owners of property are held liable for the actions of past owners despite the fact that the new owners may have no knowledge of past disposal practices or releases. Indeed, lending institutions have been held liable for these expenses and penalties under CERCLA.

While CERCLA imposed strict, joint, and several liability on PRPs, four provisions were written that would allow CERCLA liability to be avoided. These provisions allow avoidance of liability by establishing that the release or threat of release of a hazardous substance and the resulting damages were caused by an act of God, an act of war, an act or omission of a third party, or by a combination of the provisions. The third party defense was relaxed slightly under SARA to allow establishment of the innocent purchaser defense from CERCLA and other environmental liabilities. The third party defense provides a framework whereby an innocent purchaser or lender may avoid CERCLA liability by the exercise of due diligence in the investigation of a property for potential environmental problems prior to the purchase. The third party defense was further relaxed with the establishment of the *de minimis* defense, an extension of the innocent purchaser defense that can be used administratively to limit a property owner's or lender's liability prior to litigation.

Exercising due diligence constitutes making all-appropriate-inquiry into the previous ownership and uses of property in a manner that is consistent with good business practice in an effort to minimize liability. Under SARA, courts were instructed to take into account any specialized knowledge on the part of the defendant, the relationship of the purchase price to the value of the property if uncontaminated, commonly known or reasonably ascertainable information about the property, obviousness of the presence of contamination at the property, and the ability to discover any contamination by appropriate inspection. If a purchaser or lender acquires property with actual or constructive knowledge of contamination, the innocent landowner defense is not available. Lenders are further impacted as they must exercise due diligence and meet the all-appropriate-inquiry requirements at both the time of initial acquisition and at foreclosure.

The establishment of all-appropriate-inquiry is the most difficult task in establishing the innocent landowner defense. The courts have held that the mere performance of a site assessment does not constitute due diligence. Environmental site assessments must investigate for the range of potential contaminants that could reasonably be expected to occur at a site based on known and/or demonstrable historical property use. Potential environmental contaminants from recent operations on a property are generally identified with minimal effort from Material Safety Data Sheets and other information sources. It is generally much more difficult to determine the range of potential historical contaminants which could cause significant CERCLA-type liability. This book will assist in the completion and evaluation of the environmental assessment process by presenting types of contaminants and specific compounds historically associated with various types of facilities, products and processes. These data have been compiled from various chemical, occupational health, and public information sources in an attempt to provide anyone potentially impacted by CERCLA-type liability with a means to determine environmental contaminants that could be present on a property.

Chapter 2 covers general considerations for the performance of environmental site assessments.     Chapter 3 discusses a method for maximizing information gained through the use of this reference.  Chapter 4 presents environmental concerns that should be evaluated for every property.  Chapter 5 presents environmental contaminants that are known to be used in, generated in, and/or associated with specific facilities, processes, and products.     Appendix A discusses transformation mechanisms and breakdown products of common contaminants which are routinely found in the environment.  Appendix B discusses methods for assessing the mobility of environmental contaminants.  Appendix C provides a list of information sources that may be consulted for more information on specific contaminants.

## THE ENVIRONMENTAL ASSESSMENT PROCESS

### Introduction

To qualify for the protection from CERCLA-type liability under the innocent purchaser defense, all-appropriate-inquiry and due diligence must be conducted before purchasing a property. Environmental assessments are conducted with the intent to satisfy the all-appropriate-inquiry and due diligence guidelines. Key to the discussion of all-appropriate-inquiry and due diligence is the determination of the good and customary business practices associated with minimizing liability. The definition of what constitutes appropriate inquiry and due diligence is influenced by economic, political, and technological forces which will continue to evolve. At minimum, appropriate inquiry should consider the following:

1.    pertinent information regarding current and previous owners and operators of the property; and

2.    inspection of the land and buildings on the property; and

3.    pertinent information obtained from environmental and/or regulatory agencies to determine if there are or were any suspected or known contamination or compliance problems on the property.

Other activities may also be necessary depending on the transaction, the type of property involved, the historical information gathered, and the results of the property inspection. The combination of these investigative activities make up what is known as an environmental site assessment. The overall goal of the environmental site assessment is to identify real, perceived, and/or potential environmental issues which may affect the environmental integrity of the property and may represent liability to a purchaser or lender.

Environmental assessments are not always conducted solely for liability protection for a purchaser or lender. Environmental assessments requested by the owner/operator of a facility or property may focus on compliance issues associated with the facility's operations. Environmental assessments may also be used by the owner/operator of a facility or property to identify potential environmental problems. These problems may then be remediated to minimize future liability.

Environmental assessments are typically conducted in phases.  The scopes of various phases are generally designed to gain maximum information concerning potential environmental problems with minimum cost exposure.   A Phase I environmental assessment is generally designed to gather background information about a property.  A Phase I assessment does not typically include sampling although it may be prudent to include sampling and analysis of the potable water supply (whether private or municipal) in the Phase I assessment.   Phase II environmental assessments are typically designed to determine if potential liabilities identified during the Phase I assessment should indeed be causes of concern.  Phase III assessments are scoped to determine the extent of contamination verified during a Phase II assessment and the associated remediation costs.

Environmental assessments do not always progress through each of the three phases.   Assessments on properties which are found to exhibit minimal liability may be completed in a single phase.  Additional phases are generally undertaken only when properties exhibit risk of substantial contamination.

Many real estate transactions have been concluded transferring contaminated properties.  Transactions such as these may require creative and practical solutions to remediate the property and allow transfer of ownership while minimizing liability to the new owner.  The information gained during the completion of various phases of environmental assessments plays a significant role in this process.

## Phase I Environmental Assessments

A Phase I environmental assessment is performed to determine if there are actual or suspected environmental problems on or related to a property.   Numerous books and articles have been published suggesting minimum requirements for Phase I assessments.  Various organizations and individual businesses have developed, with assistance from legal counsel, their own minimum standards for Phase I assessments.

Phase I assessments typically focus on a title search and review of related documents, inspections of the property and interviews with persons who have knowledge of the property, and contacting appropriate environmental and regulatory agencies for information on the property.  Operations on surrounding properties should be reviewed to determine the potential for migration of contaminants onto the subject property.  The assessment should also consider regulatory agency records of known environmental problems on surrounding properties in the vicinity of the subject property.

## The Title Search and Review of Related Documents

CERCLA specifically refers to the investigation of historical property ownership and use as being part of all-appropriate-inquiry and due diligence Property records should be examined to determine all previous owners of the property of record. The chain of title will give the names of previous owners of record and may provide information concerning previous uses of the property. A common error that is made is to request title information only for the previous 40 years. Many current sites of known environmental contamination in the United States resulted from Civil War (1860s) vintage munitions and coal gasification facilities. The chain of title for a property should encompass a sufficiently broad time period to allow a substantially complete review of the property's ownership history. In most cases, property use is of more importance than ownership in determining the potential for environmental impairment of a property.

Many sources of information regarding historical property use are available. Aerial photographs, obtainable through the US Department of Agriculture and local taxing agencies, offer an excellent window into both the past and present use of a property and its surroundings. Much information concerning the use of a property and its surroundings can be found through local historical libraries. Newspaper articles often discuss significant historical events and activities that may have affected a site. Historical maps may also be available through local libraries and government offices that may provide information about former site uses. City directories, tax records, water and sewer processing records, local fire station records, and zoning records may also provide useful information for determining the types of potential environmental contaminants that should be investigated.

## Property Inspections and Interviews with Knowledgeable People

The most important segment of the Phase I environmental assessment involves the inspection of the subject property and interviews with persons knowledgeable about the property. The inspection allows collection of visual evidence as well as the potential verification of information obtained during historical data gathering.

Prior to performing the inspection, all buildings and areas to be inspected should delineated. The boundaries of the property and utility corridors should also be established. In large or complex industrial sites, it may be appropriate to be accompanied by a person knowledgeable about process operations and safety considerations. All areas that are not accessible during the inspection should be noted and treated as suspect. The inspection of a property should be used to ascertain the potential for environmental contamination that was uncovered through records searches and aerial photographic reviews. Inspections are often performed, however, prior to the completion of the background information collection and review due to the timing of the real estate transaction and client-imposed deadlines. This circumstance often lessens the value of the environment assessment by reducing the amount of information that the environmental consultant is able to utilize prior to going onsite.

Inspection of a property involves visually examining the property grounds, facilities and buildings at the site, surface soils, and surface waters for signs of contamination and for potential contaminant migration sources and pathways. The inspection may also include screening soils and water for indirect signs of contamination such as odor and pH. Signs of potential environmental contamination may include such things as stressed vegetation, stained or discolored soils, floors, and walls, unnatural fill materials, areas of chemical storage and handling, storage tanks and structures, discharge locations, and monitoring wells. More subtle clues indicating potential environmental contamination may include disturbed or sunken soils and multiple vintages of asphalt or concrete that may not correspond with property development history. It is also important to identify and inspect all potential migration contaminant pathways. Potential migration pathways may include such things as cracks and perforations in floors, floor drains, dry wells, septic drain fields, waste water discharges, and lagoons.

Raw materials, intermediate products, final products, and waste products should be considered during the property inspection process. Environmental contamination frequently occurs by the mismanagement of raw materials and by unsafe and/or illegal processes and disposal practices. The storage and use of raw materials may be regulated by federal, state, and/or local agencies, especially when hazardous materials are involved. Associated materials used to facilitate a process, such as cleaners and solvents, should also be considered. Materials in use at a facility can be determined from a variety of sources. These sources include Material Safety Data Sheets (MSDS) and internal purchase orders, interviews with knowledgeable personnel, onsite investigations, Occupational Health and Safety Administration (OSHA) records, Pollution Incident and Prevention Plans (PIPP), and SARA Title III - Community Right to Know databases.

Visual observations of adjacent properties for indicators of potential environmental problems should be performed. It is best to observe adjacent properties either from the subject property or from public areas such as the streets and sidewalks. Visual observations may include such things as gasoline pumps and fill pipes, stressed vegetation and stained soils, above ground tanks, drum storage areas, lagoons, poor disposal and waste handling practices, transformers, and evidence of fires. Drainage patterns should be reviewed to ascertain the likelihood of adverse environmental impact to the subject property by surface run-off from adjacent properties.

A Phase I environmental assessment does not typically include sampling. It is not generally possible to determine a sampling and analytical regimen that will satisfy the all-appropriate-inquiry and due diligence criteria (along with a reasonable cost proposal) prior to the completion of the Phase I assessment. Sampling is generally most representative and cost-effective when the number of samples and types of analyses are based on a reasonable, logical sampling plan that takes into account all information gathered during the Phase I assessment. Sampling and analytical data obtained prior to the completion and evaluation of a Phase I assessment should not be considered as complete nor representative of the range of potential contaminants that could be present and which could cause significant liability for a future owner or lender.

### Regulatory Agency Reviews

The last major element of an Phase I environmental assessment is a review of available regulatory agency records for any information which may shed light on the environmental condition of a property. Information should be gathered that indicates knowledge of contamination at or near the site. These data should include the identification of all violations, fines, penalties, and/or pending actions against the owners or operators of a property. Air and surface water discharge permits, inspection reports, waste manifests, waste summaries, and discharge monitoring reports may be helpful in evaluating compliance history and potential environmental contaminants which may be or may have been used at a facility.

A typical review of agency records begins with the searching of applicable databases which might contain information about the subject property and/or surrounding properties. A number of federal, state, and local databases are available that can provide information regarding known and potential sites of environmental contamination. These databases include the Federal Superfund Data Base - CERCLIS, the National Priority List - NPL, state priority lists, landfill and solid waste facility databases, registered underground storage tank databases, and leaking underground storage tank databases. Additional databases with useful information may include the Hazardous Waste Generator information as required under RCRA, SARA Title III, hydrocarbon contamination from oil and gas drilling, National Pollution Discharge Elimination System (NPDES) permits, air emission permits, Toxic Substances Control Act (TSCA) and Federal Insecticide, Fungicide and Rodenticide Act (FIFRA), wetlands listings, and Security and Exchange Commission 10k Reports. Computerized searches of many of these databases are available from a number of specialized commercial reporting services. Additional sources of information include EPA Point Source Category Development Documents, local health department and fire department records.

The geographical search radius for gathering information on sites of known or potential contamination surrounding a property depends on local hydrogeology and other site-specific factors. It has been suggested that, in extreme cases, it may be necessary to consider information from National Priority List sites within a four-mile radius of the subject property if large volumes of contaminants are or were present and conditions are conducive to long distance migration of contaminants. Typical radii of investigation for sites of known environmental contamination varies from one-half mile to one mile.

Information gathered during a database search may indicate that the agencies queried have no record of sites of known or potential contamination within the search area. This will generally be the case in rural and agricultural areas and in areas that have only recently developed. It may also be found, however, that a number of sites of environmental contamination fall within the search radius, necessitating sufficient investigation into agency files to ascertain the degree of adverse environmental impact on or to the property from these sites.

## Other Considerations

The various databases and information sources listed above are quite useful for obtaining information about current operations on a property or at a site. It is much more difficult to ascertain what types of potential environmental contaminants may have historically been used at a site or in a process. Many chemicals and radiation sources have half-lives that span many years. Historical associations of potential contaminants with various manufacturing processes and products follow in later sections.

## The Phase I Environmental Assessment Report

Information gathered during a Phase I environmental assessment should be presented clearly, concisely, and logically. Presentation format should be tailored to the needs of the individual or company who requested the assessment. The report should also include a conclusions section which indicates the environmental consultant's opinion of the potential for contamination based on the information gathered during Phase I activities. Phase I environmental assessment reports may or may not include a proposal for additional recommended investigative activities.

The Phase I environmental assessment report should not certify that a property is free of contamination nor liability. Likewise, the Phase I report should not guess at the extent of contamination without appropriate investigation and analyses. The report should present a basis for the prospective buyer or lender to make an informed decision whether to proceed with or cancel the transaction. If the Phase I assessment turns up no evidence of potential contamination but subsequent contamination is found on the property, the report could be used to demonstrate that all-appropriate-inquiry and due diligence were conducted prior to the closing of the transaction.

## The Phase II Environmental Assessment

Often, there are additional investigative activities that may be needed to address questions and concerns raised during the Phase I environmental assessment. These generally must be addressed prior to closing the transaction in order to satisfy the all-appropriate-inquiry and due diligence requirements of the innocent landowner defense. The scope of the Phase II environmental assessment should be designed to address the specific concerns raised by the Phase I environmental assessment.

A Phase II assessment is primarily used to sample for and ascertain the presence of specific contaminants at a site. The range of potential contaminants the Phase II assessment is designed to screen for must be representative of the compounds which have been historically associated with site use. Inappropriate or incomplete sampling or analysis may risk innocent landowner defense or *de minimis* settlement options which might otherwise be available to property owners and lenders under SARA.

Phase II activities include soil borings, soil gas surveys, soil sampling, and analysis, surface and ground water sampling, underground storage tank testing, etc. A Phase II assessment may also include the review of records and interviews or investigations of persons outside the scope of work of the Phase I assessment. This circumstance frequently occurs when price competition keeps an environmental consultant from performing a thorough review of all pertinent information.

## Additional Investigations

If environmental contamination has been confirmed during a Phase II assessment and the parties to the property transaction wish to continue pursuing the transaction, it may be necessary to perform further investigations. These investigations should be designed to determine the extent of contamination so that remedial costs may be estimated and/or cleanup undertaken. These additional investigations may also involve environmental risk assessments that consider the toxicity of contaminants, populations at risk, and the effects of short- and long-term exposures. It is important for purchasers or lenders, if they are not already represented, to obtain legal counsel experienced in environmental issues to represent their interests at this point in the transaction as significant financial and environmental liability may be incurred if investigations, risk assessments, and/or remediation are not conducted properly.

## GUIDE TO THE USE OF THIS HANDBOOK

### Introduction

The Handbook of Environmental Contaminants is designed to assist in the determination of the types of contaminants that may be associated with a property based on the documented historical uses of that property. The chemical compounds listed in the following sections are those chemicals that have been documented as being historically associated with and/or indicators of the various activities, facilities, processes, and products listed. These and other data should be evaluated when determining analytical testing programs to ascertain the presence of contaminants that may represent environmental liability. Determination of historical activities and occurrences that may have affected the property are crucial to the all-appropriate-inquiry and due diligence aspects of the innocent landowner defense.

### Determining Potential Environmental Contaminants

The first step in determining potential environmental contaminants that could be present on or resulting from a property is to obtain a Phase I environmental assessment of the property. A properly conducted Phase I assessment should determine the range of historical activities conducted at the site. As mentioned previously, these historical activities are obtained through the review of applicable public records, private records, and interviews with key individuals. The Phase I assessment should note the presence of any indicators of potential environmental contamination that were observed on or adjacent to the property and in photographs of the property. Lastly, the Phase I assessment should document historical events on the property that may indicate the potential for environmental contamination.

A second step in determining potential environmental contaminants on a property is the evaluation of potential contaminant migration pathways and health risks that are common to most every site. Contaminant migration pathways include the sewage, waste water, and storm water disposal systems, septic systems and drain fields, dry wells, and underground storage tanks. It may be prudent to evaluate private and municipal water supplies for a property or facility. Other topics which may be evaluated at every site are asbestos, electromagnetic radiation, noise, and indoor air quality.

The third step in the evaluation process is to review appropriate purchasing, storage, shipping, and disposal records for the types of products that are recorded as having been handled on the property. These records may be available only from the most recent occupant of the site but are useful in identifying potential environmental contaminants which were documented on the property.

Upon completion of these steps, it is possible to begin to determine the types of potential environmental contaminants that may be associated with a property. All known current and historical events, facility types, processes, activities, and/or products that are associated with the property should be investigated. Each of these issues should be researched to determine general and specific types and quantities of products and chemicals that may have been used and which could represent environmental liability if they were to be found as environmental contaminants on a property.

After determining the range of potential contaminants associated with a property, chemical mobility and breakdown products in the environment should be evaluated to determined the various resources that could be affected by potential contaminants. Chemical breakdown products for common solvents are discussed in Appendix A. Chemical mobility is discussed in Appendix B. Information regarding specific chemical uses and characteristics may be obtained from commonly used chemical reference books. Some additional information sources are presented in Appendix C.

## Discussion

The failure to evaluate for the full range of potential contaminants associated with a site may cause unnecessary liability to be incurred. While analytical testing for a number of potential contaminants can be somewhat costly and time-consuming, the costs of environmental liability can be much higher. Professional and/or legal counsel should be obtained to assess the potential for liability when determining if a less than exhaustive sampling and analytical program would satisfy the requirements for all-appropriate-inquiry and due diligence.

## POTENTIAL CONTAMINANTS TO INVESTIGATE
## AT EVERY DEVELOPED SITE

**Introduction**

Information collected during the Phase I environmental assessment should include specific investigation into potential contaminant sources and health-risk concerns that are present at every developed site.   Common problem areas include the potable water supply, sewage disposal, and septic systems and drain fields.   Underground storage tanks in which fuels, feed stocks and products, and waste products may be stored at sites are also common problem areas.   Indicator compounds and compounds of potential concern associated with these problem areas will be identified in this section. Other topics which should be evaluated at every site, but are not discussed in detail are asbestos, electromagnetic radiation, noise, and indoor air quality, including radon.

Ground water from private wells provides an excellent initial noninvasive assessment of the potential for significant ground water contamination at a property or facility.   It is also important to analyze municipal water supplies as increasing evidence suggests that municipal water supplies may carry compounds such as lead at levels that pose significant health risk.   The analysis of a potable water supply well or wells can be easily incorporated into a Phase I environmental assessment with minimal personnel charges or time incurred.

Sewage, septic systems, and drain fields are significant potential contaminant pathways that exist at almost every site.   These systems should be checked not only for indicators of excessive human and/or animal waste, but for any potential contaminants that may have been used at the site.

Underground storage tanks pose widely recognized risks of environmental contamination. Significant contamination has also been shown to persist from previously existing tank farms.

## EPA PRIMARY DRINKING WATER PARAMETERS

Inorganic Compounds

- Alpha particle activity
- Arsenic
- Antimony (proposed)
- Barium
- Beryllium (proposed)
- Beta particle activity
- Cadmium
- Chromium
- Cyanide (proposed)
- Fluorides
- Lead
- Mercury
- Nickel (proposed)
- Nitrates
- Radium 226 and 228
- Selenium
- Silver
- Sulfate (proposed)
- Thallium (proposed)
- Turbidity

Volatile Organic Compounds

- Benzene
- Carbon tetrachloride
- o-Dichlorobenzene
- p-Dichlorobenzene
- 1,2 Dichloroethane
- 1,1 Dichloroethene
- cis 1,2 Dichloroethene
- trans 1,2 Dichloroethene
- 1,2 Dichloropropane
- Ethyl benzene
- Methylene chloride (proposed)
- Monochlorobenzene
- Styrene
- Tetrachloroethylene
- Toluene

Volatile Organic Compounds (cont.)

- 1,1,1 Trichloroethane
- 1,1,2 Trichloroethane (proposed)
- Trichloroethene
- Vinyl chloride
- Xylenes

Base Neutral Extractables/Pesticides/Herbicides/PCBs

- 2,4-D
- Acrylamide
- Adipates (proposed)
- Alachlor
- Aldicarb
- Aldicarb sulfone
- Aldicarb sulfoxide
- Atrazine
- Carbofuran
- Dalapon (proposed)
- Dibromochloropropane
- Dinoseb (proposed)
- Diquat (proposed)
- Endothall (proposed)
- Endrin (proposed)
- Epichlorohydrin
- Ethylene dibromide
- Glyphosate (proposed)
- Heptachlor
- Heptachlor epoxide
- Hexachlorobenzene (proposed)
- Hexachlorocyclopentadiene (proposed)
- Lindane
- Methoxychlor
- Oxamyl (proposed)
- Polychlorinated biphenyls
- Phthalates (proposed)
- Picloram (proposed)
- Simazine (proposed)
- 2,3,7,8-TCDD [Dioxin] (proposed)
- Toxaphene
- 2,4,5-TP [Silvex] (proposed)
- 1,2,4 Trichlorobenzene (proposed)
- Trihalomethanes

Microbiological Factors

- Bacteria
- Coliform

Physical Factors

- Turbidity

Polynuclear Aromatic Hydrocarbons

- Benzo(a)anthracene (proposed)
- Benzo(a)pyrene (proposed)
- Benzo(b)fluoranthene (proposed)
- Benzo(k)fluoranthene (proposed)
- Chrysene (proposed)
- Dibenz(a,h)anthracene (proposed)
- Indenopyrene (proposed)

Radionuclides

- Gross alpha activity
- Man-made beta activity
- Radium 226
- Radium 228

## EPA SECONDARY DRINKING WATER PARAMETERS

- Aluminum
- Chlorides
- Color
- Copper
- Corrosivity
- o-Dichlorobenzene
- p-Dichlorobenzene
- Ethyl benzene
- Fluoride
- Foaming agents

## EPA SECONDARY DRINKING WATER PARAMETERS (cont.)

- Hexachlorocyclopentadiene (proposed)
- Iron
- Manganese
- Odor
- Pentachlorophenol
- pH
- Silver
- Styrene
- Sulfate
- Total dissolved solids
- Xylenes
- Zinc

## SEWAGE, SEWAGE TREATMENT, SEPTIC SYSTEMS, AND DRAIN FIELDS

Indicator Compounds

- Alkalinity
- Aluminum
- Ammonium nitrate
- Arsenic
- Bacteria
- Biochemical oxygen demand (BOD)
- Cadmium
- Calcium oxide
- Chemical oxygen demand (COD)
- Chlorides
- Chlorinated lime
- Chromium
- Copper
- Fecal coliform
- Fluoride
- Grease
- Hydrogen sulfide
- Iron
- Kjeldahl nitrogen

Indicator Compounds (cont.)

- Lead
- Magnesium
- Manganese
- Mercury
- Methylene blue-activated substances
- Nitrates
- Nitrites
- Pesticides
- pH
- Phosphorus
- Potassium
- Orthophosphates
- Selenium
- Semivolatile compounds
- Sludges
- Sodium
- Solids
- Sulfates
- Trihalomethanes
- Turbidity
- Volatile solids
- Viruses
- Zinc

## UNDERGROUND STORAGE TANKS

Diesel Fuel, Jet Fuel, and Fuel Oils

- Additives
- Alcohols
- Benzene
- Ethyl benzene
- Polynuclear aromatic hydrocarbons
- Toluene
- Total petroleum hydrocarbons
- Xylenes

Gasoline

- Additives
- Alcohols
- Benzene
- Ethyl benzene
- Lead
- Methyl tertiary butyl ether
- Toluene
- Xylenes

Waste Oils

- Alcohols
- Benzene
- Ethyl benzene
- Heavy metals
- Polychlorinated biphenyls
- Polynuclear aromatic hydrocarbons
- Solvents
- Toluene
- Total petroleum hydrocarbons
- Xylenes

## SITES OF HISTORICAL FIRES

General Types of Associated Materials

- Fire-extinguishing materials
- Fire retardants

Potential Contaminants Historically Used to Fight Fires

- Carbon tetrachloride
- Chloroform
- Dibromomethane
- Freon 11

Potential Contaminants Generated During Fires

- Dioxins
- Formaldehyde
- Freon 11C
- Furans
- Phosgene
- Polynuclear aromatic hydrocarbons
- Sodium bicarbonate
- Tetrahydrofurans

## FACILITY, PROCESS, AND PRODUCT-SPECIFIC CONTAMINANTS

### 2,4,5-T

General Types of Associated Materials:

- Chlorinated herbicides

Raw Materials, Intermediate Products, Final Products, and Waste Products Generated During Manufacture and Use:

- See Herbicides

### 2,4,5-TP

General Types of Associated Materials:

- Chlorinated herbicides

Raw Materials, Intermediate Products, Final Products, and Waste Products Generated During Manufacture and Use:

- See Herbicides

### 2,4-D

General Types of Associated Materials:

- Chlorinated herbicides

Raw Materials, Intermediate Products, Final Products, and Waste Products Generated During Manufacture and Use:

- See Herbicides

## 2,4-DB

General Types of Associated Materials:

- Chlorinated herbicides

Raw Materials, Intermediate Products, Final Products, and Waste Products Generated During Manufacture and Use:

- See Herbicides

## Abate

General Types of Associated Materials:

- Organophosphate insecticides

Raw Materials, Intermediate Products, Final Products, and Waste Products Generated During Manufacture and Use:

- See Insecticides

## Abrasives

General Types of Associated Materials:

- Emery
- Glues
- Resins

Raw Materials, Intermediate Products, Final Products, and Waste Products Generated During Manufacture and Use:

- Aluminum
- Antimony
- Carborundum
- Zirconium

## Acetaldehyde

Raw Materials, Intermediate Products, Final Products, and Waste Products Generated During Manufacture and Use:

- Ethyl alcohol

## Acetanilide

Raw Materials, Intermediate Products, Final Products, and Waste Products Generated During Manufacture and Use:

- Acetic acid
- Aniline

Other Associated Materials:

- Acetic anhydride

## Acetates

Raw Materials, Intermediate Products, Final Products, and Waste Products Generated During Manufacture and Use:

- Acetic acid

Other Associated Materials:

- Acetic anhydride

## Acetic Acid

Raw Materials, Intermediate Products, Final Products, and Waste Products Generated During Manufacture and Use:

- Acetaldehyde
- Formic acid
- Methyl alcohol
- Sulfuric acid

Other Associated Materials:

- Acetic anhydride

## Acetic Anhydride

Raw Materials, Intermediate Products, Final Products, and Waste Products Generated During Manufacture and Use:

- Acetaldehyde
- Acetic acid
- Ethyl alcohol

## Acetone

General Types of Associated Materials:

- Ketones

Raw Materials, Intermediate Products, Final Products, and Waste Products Generated During Manufacture and Use:

- Acetic acid
- Isopropyl alcohol

## Acetyl Cellulose

General Types of Associated Materials:

- Cellulose

Other Associated Materials:

- Dichloroethene
- Formates

## Acetyl Chloride

Raw Materials, Intermediate Products, Final Products, and Waste Products Generated During Manufacture and Use:

- Acetic acid

## Acetylene

Other Associated Materials:

- Dimethylformamide

## Acid Dipping

General Types of Associated Materials:

- Acids
- Heavy metals

Raw Materials, Intermediate Products, Final Products, and Waste Products Generated During Manufacture and Use:

- Arsine
- Hydrogen cyanide

## Acriflavine

Raw Materials, Intermediate Products, Final Products, and Waste Products Generated During Manufacture and Use:

- Acridine

## Acrolein

Raw Materials, Intermediate Products, Final Products, and Waste Products Generated During Manufacture and Use:

- Acetaldehyde
- Allyl alcohol

## Acrylates

General Types of Associated Materials:

- Plastics

Raw Materials, Intermediate Products, Final Products, and Waste Products Generated During Manufacture and Use:

- Acrolein
- beta-Propiolactone
- Hydrogen cyanide

## Acrylics

Raw Materials, Intermediate Products, Final Products, and Waste Products Generated During Manufacture and Use:

- Acrylonitrile
- Nickel

## Acrylonitrile

General Types of Associated Materials:

- Acrylics
- Nitrile rubber

Raw Materials, Intermediate Products, Final Products, and Waste Products
Generated During Manufacture and Use:

- Ethylene oxide
- Hydrogen cyanide

## Activated Carbon

Raw Materials, Intermediate Products, Final Products, and Waste Products
Generated During Manufacture and Use:

- Charcoal
- Coal tar
- Polynuclear aromatic hydrocarbons
- Phenols
- Phosphoric acid

## Adhesives

Raw Materials, Intermediate Products, Final Products, and Waste Products
Generated During Manufacture and Use:

- Benzol
- Cresols
- Dichloroethane
- Ethyl silicate
- Hydrogen sulfide
- Isocyanates
- Mercaptans
- Sulfureted hydrogen
- Trichloroethene

Other Associated Materials:

- Acrolein
- Benzene
- Carbon disulfide
- Carbon tetrachloride
- Dioxane
- Ethyl alcohol
- Ethyl benzene
- Hydrochloric acid
- Ketones
- Methyl alcohol
- Nitrobenzol
- Sulfuric acid
- Xylenes

## Adhesives, Nitrile

General Types of Associated Materials:

- Adhesives

Raw Materials, Intermediate Products, Final Products, and Waste Products Generated During Manufacture and Use:

- Acrylonitrile

## Adipic Acid

Raw Materials, Intermediate Products, Final Products, and Waste Products Generated During Manufacture and Use:

- Cyclohexane
- Cyclohexene
- Cyclopropane
- Furfural
- Methylcyclohexene

## Aerospace Components

Raw Materials, Intermediate Products, Final Products, and Waste Products Generated During Manufacture and Use:

- Graphite
- Magnesium
- Titanium

## Aerosol Propellants

General Types of Associated Materials:

- Alcohols
- Chlorinated fluorocarbons

Raw Materials, Intermediate Products, Final Products, and Waste Products Generated During Manufacture and Use:

- Chlorine
- Methylene chloride

## Agricultural Chemicals

General Types of Associated Materials:

- Fertilizers
- Herbicides
- Insecticides
- Metals
- Pesticides
- Petroleum products
- Rodenticides
- Solvents

Raw Materials, Intermediate Products, Final Products, and Waste Products Generated During Manufacture and Use:

- Arsenic
- Hydrazine and derivatives
- Mercury
- Nitrates
- Phenols
- Phosphorus
- Polybrominated biphenyls
- Polychlorinated biphenyls
- Polynuclear aromatic hydrocarbons

## Aircraft

General Types of Associated Materials:

- Adhesives and removers
- Cutting fluids
- Epoxy resins
- Flame retardants
- Hydraulic fluids
- Lubricants
- Oils
- Paints
- Plastics
- Rubber
- Solvents
- Thinners

Raw Materials, Intermediate Products, Final Products, and Waste Products Generated During Manufacture and Use:

- Benzene
- Chromates
- Chromium
- Cyanides
- Hydrofluoric acid
- Nitric acid

Other Associated Materials:

- Chromic acid
- Trichloroethene

## Aircraft Components

Raw Materials, Intermediate Products, Final Products, and Waste Products Generated During Manufacture and Use:

- Graphite
- Magnesium
- Titanium

## Aircraft Engines

Raw Materials, Intermediate Products, Final Products, and Waste Products Generated During Manufacture and Use:

- Graphite
- Magnesium
- Tin
- Titanium

## Airplane Dopes

Raw Materials, Intermediate Products, Final Products, and Waste Products Generated During Manufacture and Use:

- Formic acid

Other Associated Materials:

- Ketones

## Airports

General Types of Associated Materials:

- Aircraft
- Aviation fuels
- Cleaners
- Detergents
- Disinfectants
- Solvents

Other Associated Materials:

- Acetone
- Benzene
- Ethyl benzene
- Methyl tertiary butyl ether
- Toluene
- Trichloroethane
- Xylenes

## Albumin

Raw Materials, Intermediate Products, Final Products, and Waste Products Generated During Manufacture and Use:

- Ethylenediamine

## Alcohol, Denaturing

General Types of Associated Materials:

- Alcohols

Raw Materials, Intermediate Products, Final Products, and Waste Products Generated During Manufacture and Use:

- Ethyl ether
- Pyridine

## Alcohol, Rubbing

Raw Materials, Intermediate Products, Final Products, and Waste Products Generated During Manufacture and Use:

- Isopropyl alcohol

## Aldehydes

Raw Materials, Intermediate Products, Final Products, and Waste Products Generated During Manufacture and Use:

- Silver

## Aldol

Raw Materials, Intermediate Products, Final Products, and Waste Products Generated During Manufacture and Use:

- Acetaldehyde

## Aldrin

General Types of Associated Materials:

- Organochlorine insecticides

Raw Materials, Intermediate Products, Final Products, and Waste Products Generated During Manufacture and Use:

- See Insecticides

## Algicides

Raw Materials, Intermediate Products, Final Products, and Waste Products Generated During Manufacture and Use:

- Copper
- Sulfates

## Alkalis

Raw Materials, Intermediate Products, Final Products, and Waste Products Generated During Manufacture and Use:

- Calcium chloride
- Chlorine
- Iodine
- Manganese

Other Associated Materials:

- Carbon disulfide

## Alkali Cellulose

General Types of Associated Materials:

- Alkalis
- Cellulose

## Alkaloids

General Types of Associated Materials:

- Alkalis

Other Associated Materials:

- Ethyl ether
- n-Butyl alcohol

## Alkyd Resins

General Types of Associated Materials:

- Resins

Raw Materials, Intermediate Products, Final Products, and Waste Products
Generated During Manufacture and Use:

- Phthalic anhydride

## Alkyl Resins

General Types of Associated Materials:

- Resins

Raw Materials, Intermediate Products, Final Products, and Waste Products Generated During Manufacture and Use:

- Nitroparaffins

## Alloys

Raw Materials, Intermediate Products, Final Products, and Waste Products Generated During Manufacture and Use:

- Cerium
- Cobalt
- Germanium
- Magnesium
- Platinum
- Silver

## Allyl Alcohol

Raw Materials, Intermediate Products, Final Products, and Waste Products Generated During Manufacture and Use:

- Formic acid

## Allyl Esters

Raw Materials, Intermediate Products, Final Products, and Waste Products Generated During Manufacture and Use:

- Allyl alcohol

## alpha-BHC

General Types of Associated Materials:

- Organochlorine insecticides

Raw Materials, Intermediate Products, Final Products, and Waste Products Generated During Manufacture and Use:

- See Insecticides

## alpha-Naphthol

Raw Materials, Intermediate Products, Final Products, and Waste Products Generated During Manufacture and Use:

- alpha-Naphthylamine

## alpha-Naphthylamine

Raw Materials, Intermediate Products, Final Products, and Waste Products Generated During Manufacture and Use:

- beta-Naphthylamine

## alpha-Naphthylthiouria

Raw Materials, Intermediate Products, Final Products, and Waste Products Generated During Manufacture and Use:

- alpha-Naphthylamine

## Aluminum

Raw Materials, Intermediate Products, Final Products, and Waste Products Generated During Manufacture and Use:

- Ammonia
- Hydrogen fluoride

Other Associated Materials:

- Fluorides

## Amalgam

Raw Materials, Intermediate Products, Final Products, and Waste Products Generated During Manufacture and Use:

- Cadmium
- Mercury
- Silver

## Amines

Raw Materials, Intermediate Products, Final Products, and Waste Products Generated During Manufacture and Use:

- Dimethyl sulfate
- Ethylene chlorohydrin
- Phenols

## Aminodiphenyl

Raw Materials, Intermediate Products, Final Products, and Waste Products Generated During Manufacture and Use:

- Nitrobiphenyl

## Ammonia

Raw Materials, Intermediate Products, Final Products, and Waste Products Generated During Manufacture and Use:

- Calcium cyanamide
- Cerium
- Nitrates

## Ammonium Salts

General Types of Associated Materials:

- Ammonia

Raw Materials, Intermediate Products, Final Products, and Waste Products Generated During Manufacture and Use:

- Carbon disulfide
- Hydrogen cyanide

## Ammonium Sulfate

General Types of Associated Materials:

- Ammonia

Raw Materials, Intermediate Products, Final Products, and Waste Products Generated During Manufacture and Use:

- Sulfuric acid

## Ammunition

Raw Materials, Intermediate Products, Final Products, and Waste Products Generated During Manufacture and Use:

- Aluminum
- Antimony
- Arsenic
- Magnesium
- Phosphorus
- Trinitrotoluene

Other Associated Materials:

- Trichloroethene

## Amyl Acetate

Raw Materials, Intermediate Products, Final Products, and Waste Products Generated During Manufacture and Use:

- Amyl alcohol

## Amyl Nitrate

Raw Materials, Intermediate Products, Final Products, and Waste Products Generated During Manufacture and Use:

- Amyl alcohol

## Anesthetics

Raw Materials, Intermediate Products, Final Products, and Waste Products Generated During Manufacture and Use:

- Chloroform
- Cyclopropane
- Ethyl chloride
- Trichloroethene

Other Associated Materials:

- Methylene chloride

## Aniline Compounds

Raw Materials, Intermediate Products, Final Products, and Waste Products Generated During Manufacture and Use:

- Arsine

## Animal Feeds

Raw Materials, Intermediate Products, Final Products, and Waste Products Generated During Manufacture and Use:

- Phosphoric acid

## Animal Handling

General Types of Associated Materials:

- Bacteria
- Cleaners
- Deodorants
- Detergents
- Feeds, treated
- Germicides
- Insecticides
- Pesticides
- Viruses

## Annealing

Raw Materials, Intermediate Products, Final Products, and Waste Products Generated During Manufacture and Use:

- Ammonia
- Heavy metals

Other Associated Materials:

- Hydrogen chloride

## Anodizing

Raw Materials, Intermediate Products, Final Products, and Waste Products Generated During Manufacture and Use:

- Chromium
- Nickel

## Anthelmintics

Raw Materials, Intermediate Products, Final Products, and Waste Products Generated During Manufacture and Use:

- Tetrachloroethene
- Tin

## Anthranilic Compounds

Raw Materials, Intermediate Products, Final Products, and Waste Products Generated During Manufacture and Use:

- Naphthalene

## Anticorrosion Additives

Raw Materials, Intermediate Products, Final Products, and Waste Products Generated During Manufacture and Use:

- Hydrazine and derivatives

## Antifreeze

Raw Materials, Intermediate Products, Final Products, and Waste Products Generated During Manufacture and Use:

- Ethyl alcohol
- Ethylene glycol
- Isopropyl alcohol
- Methyl alcohol

## Antiknock Compounds

Raw Materials, Intermediate Products, Final Products, and Waste Products Generated During Manufacture and Use:

- Dichloroethane
- Dibromoethane
- Dibromomethane
- Lead, alkyl
- Lead, tetra ethyl
- Methyl tertiary butyl ether

## Antimitotic Agents

Raw Materials, Intermediate Products, Final Products, and Waste Products Generated During Manufacture and Use:

- Hydroquinone

## Antioxidants

Raw Materials, Intermediate Products, Final Products, and Waste Products Generated During Manufacture and Use:

- Acetaldehyde
- Aniline
- beta-Naphthylamine
- Cresol
- Hydrazine and derivatives
- Hydroquinone
- Nitrosodimethylamine

## Antiseptics

Raw Materials, Intermediate Products, Final Products, and Waste Product:
Generated During Manufacture and Use:

- Boron
- Magnesium
- Mercury
- Tetramethylthiuram disulfide

## Appliances, Household

General Types of Associated Materials:

- Acids
- Enamels
- Epoxy resins
- Fluxes
- Phenolic resins
- Pitch
- Rubber
- Solders
- Solvents
- Varnishes
- Waxes

Raw Materials, Intermediate Products, Final Products, and Waste Products
Generated During Manufacture and Use:

- Asbestos
- Chlorinated diphenyls
- Chromium
- Cyanides
- MOCA
- Naphthalenes, chlorinated
- Polychlorinated biphenyls
- Zinc

Other Associated Materials:

- Trichloroethene


## Arrack

Other Associated Materials:

- Formates


## Arsenic Acid

Raw Materials, Intermediate Products, Final Products, and Waste Products
Generated During Manufacture and Use:

- Arsenic
- Nitrogen


## Asbestos

Raw Materials, Intermediate Products, Final Products, and Waste Products
Generated During Manufacture and Use:

- Benzene
- Phenols


## Asphalt

General Types of Associated Materials:

- Petroleum hydrocarbons
- Solvents
- Tar

Raw Materials, Intermediate Products, Final Products, and Waste Product:
Generated During Manufacture and Use:

- Copper
- Creosote
- Dichloroethane
- Ethylenediamine
- Phenols
- Polynuclear aromatic hydrocarbons

## Aspirin

Raw Materials, Intermediate Products, Final Products, and Waste Product:
Generated During Manufacture and Use:

- Acetic acid

Other Associated Materials:

- Acetic anhydride

## Athletic Facilities

General Types of Associated Materials:

- Adhesives and removers
- Deodorants
- Oils
- Petroleum fuels
- Soaps

Raw Materials, Intermediate Products, Final Products, and Waste Product:
Generated During Manufacture and Use:

- Bacteria
- Liniments

## Atomic Weapons

General Types of Associated Materials:

- Explosives
- Radioactive materials

Raw Materials, Intermediate Products, Final Products, and Waste Products Generated During Manufacture and Use:

- Beryllium
- Plutonium
- Uranium

## Aurinacrine

Raw Materials, Intermediate Products, Final Products, and Waste Products Generated During Manufacture and Use:

- Acridine

## Automotive Assembly and Repair

General Types of Associated Materials:

- Abrasives
- Acids
- Adhesives and removers
- Alkalis
- Antifreeze
- Battery acids
- Brake fluids
- Brake linings
- Cleaners
- Detergents
- Diesel fuel
- Epoxy resins

General Types of Associated Materials (cont.):

- Flame retardants
- Gasoline
- Gasoline additives
- Heating oil
- Hydraulic fluids
- Lubricants
- Oils
- Paint removers
- Paint thinners
- Paints
- Petroleum fuels
- Rubber
- Solders
- Solvents
- Thinners
- Transmission fluids
- Waste oils

Raw Materials, Intermediate Products, Final Products, and Waste Products Generated During Manufacture and Use:

- Asbestos
- Heavy metals
- Hydrochloric acid
- Lead
- Tin

Other Associated Materials:

- Benzene
- Bromodichloromethane
- Bromoform
- Butane
- Carbon tetrachloride
- Dichlorobenzene
- Dichloroethane
- Dichloroethene
- Dichloromethane
- Ethyl benzene

Other Associated Materials (cont.):

- Hexane
- Methyl tertiary butyl ether
- Naphthalene
- Perchloroethene
- Toluene
- Trichloroethane
- Trichloroethene
- Xylenes

## Automotive Bearings

Raw Materials, Intermediate Products, Final Products, and Waste Products Generated During Manufacture and Use:

- Antimony
- Chromium
- Silver
- Tin

Other Associated Materials:

- Lubricants
- Polychlorinated biphenyls

## Automotive Finishes

General Types of Associated Materials:

- Dewaxers
- Paint
- Solvents

Raw Materials, Intermediate Products, Final Products, and Waste Products Generated During Manufacture and Use:

- Phthalic anhydride

## Automobile Plastics

Raw Materials, Intermediate Products, Final Products, and Waste Product Generated During Manufacture and Use:

- MOCA

## Aviation Fuels

Raw Materials, Intermediate Products, Final Products, and Waste Product Generated During Manufacture and Use:

- Benzene
- Ethyl benzene
- Methyl tertiary butyl ether
- Polynuclear aromatic hydrocarbons
- Toluene
- Xylenes

## Azine

Raw Materials, Intermediate Products, Final Products, and Waste Product Generated During Manufacture and Use:

- Pyridine

## Azinphos Methyl

General Types of Associated Materials:

- Organophosphorus insecticides

Raw Materials, Intermediate Products, Final Products, and Waste Product Generated During Manufacture and Use:

- See Insecticides

### Babbitt Metal

Raw Materials, Intermediate Products, Final Products, and Waste Products Generated During Manufacture and Use:

- Antimony
- Arsenic
- Tin

Other Associated Materials:

- Hydrogen chloride

### Bactericides

Raw Materials, Intermediate Products, Final Products, and Waste Products Generated During Manufacture and Use:

- Mercury
- Silver

### Bacteriostats

Raw Materials, Intermediate Products, Final Products, and Waste Products Generated During Manufacture and Use:

- Hydroquinone
- Tetramethylthiuram disulfide

### Bakelite

Raw Materials, Intermediate Products, Final Products, and Waste Products Generated During Manufacture and Use:

- Dichloroethane
- Phenols

## Bakeries

Raw Materials, Intermediate Products, Final Products, and Waste Products
Generated During Manufacture and Use:

- Benzoyl peroxide

## Barber and Beauty Shops

General Types of Associated Materials:

- Antiseptics
- Cosmetics
- Depilatories
- Detergents
- Dyes
- Hair conditioners
- Hair sprays
- Hair straighteners
- Hair tonics
- Perfumes
- Shampoos
- Shaving cream
- Wave solutions

Raw Materials, Intermediate Products, Final Products, and Waste Products
Generated During Manufacture and Use:

- Bacteria

Other Associated Materials:

- Ammonium compounds

## Barium Sulfate

Raw Materials, Intermediate Products, Final Products, and Waste Products
Generated During Manufacture and Use:

- Barium
- Sulfuric acid

## Barometers

Raw Materials, Intermediate Products, Final Products, and Waste Products
Generated During Manufacture and Use:

- Mercury

## Barrel Reclamation

General Types of Associated Materials:

- Detergents
- Solvents

Other Associated Materials:

- Acetone
- Phthalates
- Trichloroethane
- Trichloroethene

## Bars, Clubs, and Lounges

General Types of Associated Materials:

- Detergents
- Disinfectants
- Soaps

## Batteries

General Types of Associated Materials:

- Alkalis
- Epoxy resins
- Pitch
- Plastics
- Solvents

Raw Materials, Intermediate Products, Final Products, and Waste Product; Generated During Manufacture and Use:

- Antimony
- Cadmium
- Cobalt
- Copper
- Lead
- Magnesium
- Manganese
- Mercury
- Nickel
- Picric acid
- Sulfuric acid
- Zinc

Other Associated Materials:

- Benzene
- Hydrogen chloride

## Baygon

General Types of Associated Materials:

- Carbamate insecticides

Raw Materials, Intermediate Products, Final Products, and Waste Products Generated During Manufacture and Use:

- See Insecticides

## Baytex

General Types of Associated Materials:

- Organophosphate insecticides

Raw Materials, Intermediate Products, Final Products, and Waste Products Generated During Manufacture and Use:

- See Insecticides

## Bearings

General Types of Associated Materials:

- Heavy metals
- Lubricants
- Solvents

Raw Materials, Intermediate Products, Final Products, and Waste Products Generated During Manufacture and Use:

- Antimony
- Chromium
- Silver
- Tin

Other Associated Materials:

- Polychlorinated biphenyls
- Perchloroethene
- Trichloroethane

## Beet Fermentation

Raw Materials, Intermediate Products, Final Products, and Waste Products Generated During Manufacture and Use:

- Amyl alcohol

## Belt Dressing

General Types of Associated Materials:

- Turpentine

Raw Materials, Intermediate Products, Final Products, and Waste Products Generated During Manufacture and Use:

- Polynuclear aromatic hydrocarbons

## Benzal Chloride

Raw Materials, Intermediate Products, Final Products, and Waste Products Generated During Manufacture and Use:

- Benzyl chloride

## Benzaldehyde

Raw Materials, Intermediate Products, Final Products, and Waste Products Generated During Manufacture and Use:

- Benzyl chloride

## Benzene

Raw Materials, Intermediate Products, Final Products, and Waste Products
Generated During Manufacture and Use:

- Cyclohexane
- Cyclohexene
- Cyclopropane
- Methylcyclohexene
- Phenols
- Toluene

## Benzene Hexachloride

Raw Materials, Intermediate Products, Final Products, and Waste Products
Generated During Manufacture and Use:

- Benzene

## Benzoflavine

Raw Materials, Intermediate Products, Final Products, and Waste Products
Generated During Manufacture and Use:

- Acridine

## Benzoic Acid

Raw Materials, Intermediate Products, Final Products, and Waste Products
Generated During Manufacture and Use:

- Phthalic anhydride
- Toluene
- Xylenes

## Benzyl

Raw Materials, Intermediate Products, Final Products, and Waste Product
Generated During Manufacture and Use:

- Toluene

## Benzyl Alcohol

Raw Materials, Intermediate Products, Final Products, and Waste Product
Generated During Manufacture and Use:

- Benzyl chloride

## Benzyl Derivatives

Raw Materials, Intermediate Products, Final Products, and Waste Product
Generated During Manufacture and Use:

- Toluene

## beta-BHC

General Types of Associated Materials:

- Organochlorine insecticides

Raw Materials, Intermediate Products, Final Products, and Waste Product
Generated During Manufacture and Use:

- See Insecticides

## beta-Naphthol

Raw Materials, Intermediate Products, Final Products, and Waste Products Generated During Manufacture and Use:

- Naphthalene

## Beverages

Raw Materials, Intermediate Products, Final Products, and Waste Products Generated During Manufacture and Use:

- Phosphoric acid

## Biochemical Laboratories

Raw Materials, Intermediate Products, Final Products, and Waste Products Generated During Manufacture and Use:

- Benzidine

## Biological Laboratories

Raw Materials, Intermediate Products, Final Products, and Waste Products Generated During Manufacture and Use:

- Carbolic acid
- Formaldehyde
- Tetrachloroethane

## Bismuth Refining

Raw Materials, Intermediate Products, Final Products, and Waste Product Generated During Manufacture and Use:

- Tellurium

## Bitumen

Raw Materials, Intermediate Products, Final Products, and Waste Product Generated During Manufacture and Use:

- Dichloroethane
- Polynuclear aromatic hydrocarbons

Other Associated Materials:

- Methylene chloride

## Blast Furnaces

Raw Materials, Intermediate Products, Final Products, and Waste Product Generated During Manufacture and Use:

- Dioxins
- Hydrogen cyanide
- Tetrahydrofurans

## Blasting Gelatin

Raw Materials, Intermediate Products, Final Products, and Waste Product Generated During Manufacture and Use:

- Ethylene glycol dinitrate
- Nitroglycerin

## Bleaches and Bleaching

General Types of Associated Materials:

- Alkalis
- Solvents

Raw Materials, Intermediate Products, Final Products, and Waste Products Generated During Manufacture and Use:

- Bromine
- Borax
- Chlorinated lime
- Chlorine
- Fluorine
- Hydrochloric acid
- Hydrogen chloride
- Hydrogen peroxide
- Manganese
- Oxalic acid
- Potassium hydroxide
- Sodium hydroxide

## Boiler Solutions

General Types of Associated Materials:

- Acids
- Algicides
- Molluscicides

Raw Materials, Intermediate Products, Final Products, and Waste Products Generated During Manufacture and Use:

- Hydrazine and derivatives
- Sulfur dioxide

## Bolstar

General Types of Associated Materials:

- Organophosphorus insecticides

Raw Materials, Intermediate Products, Final Products, and Waste Produc'
Generated During Manufacture and Use:

- See Insecticides

## Bombs

Raw Materials, Intermediate Products, Final Products, and Waste Produc'
Generated During Manufacture and Use:

- Trinitrotoluene

## Bookbinding

General Types of Associated Materials:

- Glues
- Inks
- Resins
- Shellac
- Solvents

Raw Materials, Intermediate Products, Final Products, and Waste Produc'
Generated During Manufacture and Use:

- Formalin
- Methyl alcohol

## Borates

Raw Materials, Intermediate Products, Final Products, and Waste Products
Generated During Manufacture and Use:

- Boron

## Brake Fluid

General Types of Associated Materials:

- Petroleum hydrocarbons

Raw Materials, Intermediate Products, Final Products, and Waste Products
Generated During Manufacture and Use:

- Amyl alcohol
- Ethylene glycol
- Ketones
- Phosphorus
- Polychlorinated biphenyls
- Tricresyl phosphates

## Brake Linings

General Types of Associated Materials:

- Metals
- Resins
- Solvents

Raw Materials, Intermediate Products, Final Products, and Waste Products
Generated During Manufacture and Use:

- Asbestos
- Graphite
- Phenols
- Tin

Other Associated Materials:

- Acetone
- Benzene
- Tetrachloroethene
- Trichloroethane
- Trichloroethene

## Brass

General Types of Associated Materials:

- Metals
- Solvents

Raw Materials, Intermediate Products, Final Products, and Waste Product Generated During Manufacture and Use:

- Arsenic
- Lead
- Tin
- Zinc

## Brazing

Raw Materials, Intermediate Products, Final Products, and Waste Product Generated During Manufacture and Use:

- Acetylene
- Nitrogen

## Brazing Alloys

General Types of Associated Materials:

- Metals

Raw Materials, Intermediate Products, Final Products, and Waste Products Generated During Manufacture and Use:

- Magnesium
- Silver
- Tin

## Breweries

General Types of Associated Materials:

- Alcohols
- Bacteria

Raw Materials, Intermediate Products, Final Products, and Waste Products Generated During Manufacture and Use:

- Arsenic
- Ethyl alcohol
- Hydrogen fluoride
- Hydrogen sulfide
- Sulfur dioxide

## Brick and Masonry Works

General Types of Associated Materials:

- Epoxy resins

Raw Materials, Intermediate Products, Final Products, and Waste Product Generated During Manufacture and Use:

- Calcium oxide
- Chromates
- Ethyl silicate
- Portland cement

## Brick Cleaning

Raw Materials, Intermediate Products, Final Products, and Waste Product Generated During Manufacture and Use:

- Hydrogen fluoride
- Muriatic acid

## Brick Oil

Raw Materials, Intermediate Products, Final Products, and Waste Product Generated During Manufacture and Use:

- Creosote
- Phenols
- Polynuclear aromatic hydrocarbons

## Brines

Raw Materials, Intermediate Products, Final Products, and Waste Product Generated During Manufacture and Use:

- Chlorides
- Magnesium
- Potassium
- Sodium

## Britannia Metal

Raw Materials, Intermediate Products, Final Products, and Waste Products Generated During Manufacture and Use:

- Antimony
- Tin

## Bromides

General Types of Associated Materials:

- Brines

Raw Materials, Intermediate Products, Final Products, and Waste Products Generated During Manufacture and Use:

- Aniline
- Bromine
- Dibromoethane

## Bromoiodide Crystals

Raw Materials, Intermediate Products, Final Products, and Waste Products Generated During Manufacture and Use:

- Bromine
- Iodine
- Thallium

## Bronze

Raw Materials, Intermediate Products, Final Products, and Waste Product Generated During Manufacture and Use:

- Arsenic
- Arsine
- Copper
- Lead
- Phosphorus
- Tin

## Bronzing

General Types of Associated Materials:

- Lacquers
- Resins
- Turpentine
- Varnishes

Raw Materials, Intermediate Products, Final Products, and Waste Product Generated During Manufacture and Use:

- Antimony
- Arsenic
- Arsine
- Cyanides
- Hydrochloric acid
- Mercury
- Phosphorus
- Sodium hydroxide
- Sulfur dioxide

Other Associated Materials:

- Acetone
- Ammonia
- Ammonium compounds

Other Associated Materials (cont.):

- Amyl acetate
- Benzene
- Benzol
- Methyl alcohol

## Broom Manufacturing

General Types of Associated Materials:

- Bleaches
- Dyes
- Glues
- Pitch
- Plastics
- Resins
- Rubber
- Shellac
- Solvents
- Tar and derivatives
- Varnishes

Raw Materials, Intermediate Products, Final Products, and Waste Products Generated During Manufacture and Use:

- Bacteria

## Brush Manufacturing

- See Broom Manufacturing

## Buranol

Raw Materials, Intermediate Products, Final Products, and Waste Product Generated During Manufacture and Use:

- Acetaldehyde

## Burnishing

General Types of Associated Materials:

- Abrasives
- Polishes
- Solvents

Other Associated Materials:

- Benzene

## Bursting Charges

General Types of Associated Materials:

- Explosives

Raw Materials, Intermediate Products, Final Products, and Waste Product Generated During Manufacture and Use:

- Trinitrotoluene

## Butadiene

Raw Materials, Intermediate Products, Final Products, and Waste Products Generated During Manufacture and Use:

- Ethyl alcohol
- Furfural

Other Associated Materials:

- Dimethylformamide

## Butchering

General Types of Associated Materials:

- Brine
- Detergents
- Enzymes

Raw Materials, Intermediate Products, Final Products, and Waste Products Generated During Manufacture and Use:

- Bacteria

## Button Manufacturing

General Types of Associated Materials:

- Animal products
- Dyes
- Plastics
- Wood products

Raw Materials, Intermediate Products, Final Products, and Waste Product Generated During Manufacture and Use:

- Bacteria
- Hydrogen peroxide

## Butyl Cellosolve

Raw Materials, Intermediate Products, Final Products, and Waste Product Generated During Manufacture and Use:

- Ethylene oxide

## Butyral

Raw Materials, Intermediate Products, Final Products, and Waste Product Generated During Manufacture and Use:

- n-Propyl alcohol

## Cabbage Processing

Raw Materials, Intermediate Products, Final Products, and Waste Product Generated During Manufacture and Use:

- Dibromomethane

## Cabinet Making

General Types of Associated Materials:

- Abrasives
- Bleaches
- Cement, contact
- Glues
- Insulation agents
- Oils
- Paints
- Pigments
- Polishes
- Resins
- Shellac
- Solvents
- Stains
- Thinners

Raw Materials, Intermediate Products, Final Products, and Waste Products Generated During Manufacture and Use:

- Rosin

## Cable Sheathing

General Types of Associated Materials:

- Dyes
- Metals
- Plastics
- Resins
- Rubber
- Solvents

Raw Materials, Intermediate Products, Final Products, and Waste Products Generated During Manufacture and Use:

- Antimony

## Cable Splicing

General Types of Associated Materials:

- Epoxy resins
- Solvents

Raw Materials, Intermediate Products, Final Products, and Waste Products Generated During Manufacture and Use:

- Chlorinated diphenyls
- Chlorinated naphthalenes
- Polychlorinated biphenyls

## Cables

General Types of Associated Materials:

- Lubricants
- Metals
- Polynuclear aromatic hydrocarbons
- Solvents

Raw Materials, Intermediate Products, Final Products, and Waste Products Generated During Manufacture and Use:

- beta-Naphthylamine
- Chlorodiphenyls

## Caffeine Processing

Raw Materials, Intermediate Products, Final Products, and Waste Products Generated During Manufacture and Use:

- Trichloroethane

### Calamine Lotion

Raw Materials, Intermediate Products, Final Products, and Waste Products Generated During Manufacture and Use:

- Zinc

### Calcium Cyanide

Raw Materials, Intermediate Products, Final Products, and Waste Products Generated During Manufacture and Use:

- Calcium cyanamide

### Camphor

General Types of Associated Materials:

- Polynuclear aromatic hydrocarbons
- Turpentine

Raw Materials, Intermediate Products, Final Products, and Waste Products Generated During Manufacture and Use:

- Naphthalene
- Trichloroethane

Other Associated Materials:

- Carbon disulfide

## Candle Manufacturing

General Types of Associated Materials:

- Paraffins
- Solvents
- Waxes

Raw Materials, Intermediate Products, Final Products, and Waste Product
Generated During Manufacture and Use:

- Chromates
- Hydrochloric acid
- Potassium nitrate
- Sodium hydroxide
- Stearic acid

Other Associated Materials:

- Ammonium compounds
- Borax
- Boric acid
- Chlorine

## Candy Makers

General Types of Associated Materials:

- Dyes
- Pesticides
- Waxes

Raw Materials, Intermediate Products, Final Products, and Waste Product
Generated During Manufacture and Use:

- Tartaric acid

## Cane

Raw Materials, Intermediate Products, Final Products, and Waste Products Generated During Manufacture and Use:

- Fluorides

## Canning, Food and Beverages

General Types of Associated Materials:

- Bacteria
- Brines
- Metals
- Preservatives
- Solders
- Solvents

Raw Materials, Intermediate Products, Final Products, and Waste Products Generated During Manufacture and Use:

- Aluminum
- Lead
- Tin

## Car Dealerships

- See Automotive Dealer

## Carbanilide

Raw Materials, Intermediate Products, Final Products, and Waste Products Generated During Manufacture and Use:

- Carbon disulfide

## Carbaryl

General Types of Associated Materials:

* Carbamate insecticides

Raw Materials, Intermediate Products, Final Products, and Waste Product Generated During Manufacture and Use:

* See Insecticides

## Carbides

Raw Materials, Intermediate Products, Final Products, and Waste Product Generated During Manufacture and Use:

* Carbon
* Tellurium
* Titanium
* Tungsten

## Carbitols

Raw Materials, Intermediate Products, Final Products, and Waste Product Generated During Manufacture and Use:

* Ethylene chlorohydrin

## Carbolic Acid

Raw Materials, Intermediate Products, Final Products, and Waste Product Generated During Manufacture and Use:

* Benzene
* Dichloroethene
* Phenols

### Carbon Tetrachloride

Raw Materials, Intermediate Products, Final Products, and Waste Products Generated During Manufacture and Use:

- Carbon disulfide
- Sulfur chloride

### Carpentry

- See Wood Working, Cabinet Making

### Carpet Manufacturing

General Types of Associated Materials:

- Bleaches
- Fungicides
- Glues
- Insecticides
- Lubricants
- Solvents

Raw Materials, Intermediate Products, Final Products, and Waste Products Generated During Manufacture and Use:

- Alizarine dyes
- Aniline dyes
- Anthrax bacillus
- Boron

Other Associated Materials:

- Chlorine

## Case Hardening

General Types of Associated Materials:

- Metals
- Oils

Raw Materials, Intermediate Products, Final Products, and Waste Product Generated During Manufacture and Use:

- Chromium
- Cyanides
- Sodium carbonate
- Sodium nitrite

## Casein

Raw Materials, Intermediate Products, Final Products, and Waste Product Generated During Manufacture and Use:

- Ethylenediamine

## Castings

- See Foundries

## Catalytic Converters

Raw Materials, Intermediate Products, Final Products, and Waste Product Generated During Manufacture and Use:

- Platinum

## Catgut

Raw Materials, Intermediate Products, Final Products, and Waste Products Generated During Manufacture and Use:

- Chloroform

## Cathode Ray Tubes

Raw Materials, Intermediate Products, Final Products, and Waste Products Generated During Manufacture and Use:

- Beryllium
- Graphite

## Cathodes

Raw Materials, Intermediate Products, Final Products, and Waste Products Generated During Manufacture and Use:

- Carbon
- Mercury

## Caustic Soda

Raw Materials, Intermediate Products, Final Products, and Waste Products Generated During Manufacture and Use:

- Mercury
- Sodium hydroxide

## Cellophane

General Types of Associated Materials:

- Plastics
- Resins
- Solvents
- Viscose

Raw Materials, Intermediate Products, Final Products, and Waste Product Generated During Manufacture and Use:

- Carbon bisulfide
- Hydrogen sulfide
- Sodium hydroxide

Other Associated Materials:

- Ethylene glycol ether

## Celluloids

Raw Materials, Intermediate Products, Final Products, and Waste Product Generated During Manufacture and Use:

- Dibromomethane
- Dinitrobenzene
- Epichlorohydrin
- Ketones
- Naphthalene
- Oxalic acid

Other Associated Materials:

- Formates
- Methyl alcohol

## Cellulose

General Types of Associated Materials:

- Acetates
- Acids
- Alkalis
- Bleaches
- Oils

Raw Materials, Intermediate Products, Final Products, and Waste Products Generated During Manufacture and Use:

- Hydrogen cyanide
- Hydrogen fluoride
- Nitrogen
- Phosphorus
- Propylene dichloride
- Sulfuric acid

Other Associated Materials:

- Carbon disulfide
- Copper
- Dichloroethyl ether
- Formates
- Nitroparaffins

## Cellulose Acetate

General Types of Associated Materials:

- Acetates
- Cellulose

Raw Materials, Intermediate Products, Final Products, and Waste Product Generated During Manufacture and Use:

- Chlorinated benzenes
- Dichloroethane
- Ethyl benzene
- Phthalic anhydride

Other Associated Materials:

- Acetic acid
- Acetic anhydride
- Dioxane
- Ethylene chlorohydrin
- Formates
- Furfural
- Ketones

## Cellulose Esters

General Types of Associated Materials:

- Cellulose

Raw Materials, Intermediate Products, Final Products, and Waste Product Generated During Manufacture and Use:

- Epichlorohydrin
- Formaldehyde
- Trichloroethane

Other Associated Materials:

- Ethylene chlorohydrin
- Ketones
- Tetrachloroethene

### Cellulose Ethers

Raw Materials, Intermediate Products, Final Products, and Waste Products Generated During Manufacture and Use:

- Epichlorohydrin

### Cellulose Formate

General Types of Associated Materials:

- Cellulose
- Cellulose esters

Raw Materials, Intermediate Products, Final Products, and Waste Products Generated During Manufacture and Use:

- Formic acid

### Cements, Rubber and Plastic

General Types of Associated Materials:

- Epoxy resins
- Pitch
- Resins
- Solvents

Raw Materials, Intermediate Products, Final Products, and Waste Products Generated During Manufacture and Use:

- Boron
- Carbon disulfide
- Cobalt
- Ethanolamines
- Tetrachloroethane
- Xylenes

## Ceramics Manufacturing

General Types of Associated Materials:

- Metals
- Solvents

Raw Materials, Intermediate Products, Final Products, and Waste Product Generated During Manufacture and Use:

- Acetone
- Antimony
- Arsenic
- Beryllium
- Cobalt
- Fluorides
- Hydrogen fluoride
- Lead
- Manganese
- Molybdenum
- Oxalic acid
- Phosphoric acid
- Platinum
- Tetrachloroethene
- Thorium
- Tin
- Titanium
- Trichloroethane
- Trichloroethene
- Uranium
- Vanadium
- Zinc
- Zirconium

## Charcoal Manufacturing

General Types of Associated Materials:

- Coal tar products

Raw Materials, Intermediate Products, Final Products, and Waste Products Generated During Manufacture and Use:

- Carbon
- Creosote
- Phenols
- Polynuclear aromatic hydrocarbons

## Chemical Indicators

Raw Materials, Intermediate Products, Final Products, and Waste Products Generated During Manufacture and Use:

- Dinitrophenol
- Nitrophenol
- Tetryl

## Chemical Weapons

Raw Materials, Intermediate Products, Final Products, and Waste Products Generated During Manufacture and Use:

- Bromine
- Chlorine

## China

Raw Materials, Intermediate Products, Final Products, and Waste Products Generated During Manufacture and Use:

- Aluminum

## Chloral

Raw Materials, Intermediate Products, Final Products, and Waste Product Generated During Manufacture and Use:

• Acetaldehyde

## Chlordane

General Types of Associated Materials:

• Organochlorine insecticides

Raw Materials, Intermediate Products, Final Products, and Waste Product Generated During Manufacture and Use:

• See Insecticides

## Chlorinated Benzenes

Raw Materials, Intermediate Products, Final Products, and Waste Product Generated During Manufacture and Use:

• Benzene

## Chlorinated Lime

Raw Materials, Intermediate Products, Final Products, and Waste Product Generated During Manufacture and Use:

• Chlorine

## Chlorinated Naphthalenes

Raw Materials, Intermediate Products, Final Products, and Waste Products Generated During Manufacture and Use:

- Chlorodiphenyls
- Naphthalene

## Chlorinated Solvents

Raw Materials, Intermediate Products, Final Products, and Waste Products Generated During Manufacture and Use:

- Chlorine
- Petroleum hydrocarbons

## Chlorination Agents

Raw Materials, Intermediate Products, Final Products, and Waste Products Generated During Manufacture and Use:

- Chlorinated lime
- Sulfur chloride

## Chlorine

Other Associated Materials:

- Hydrogen chloride

## Chloroacetic Acid

Raw Materials, Intermediate Products, Final Products, and Waste Product Generated During Manufacture and Use:

- Trichloroethene

## Chloroecholine Chloride

Raw Materials, Intermediate Products, Final Products, and Waste Product Generated During Manufacture and Use:

- Trichloroethane

## Chloroform

Raw Materials, Intermediate Products, Final Products, and Waste Product Generated During Manufacture and Use:

- Chlorine
- Formalin
- Nitrogen

## Chlorpyrifos

General Types of Associated Materials:

- Organophosphorus insecticides

Raw Materials, Intermediate Products, Final Products, and Waste Product Generated During Manufacture and Use:

- See Insecticides

### Chrysaniline

Raw Materials, Intermediate Products, Final Products, and Waste Products Generated During Manufacture and Use:

- Acridine

### Citric Acid

Raw Materials, Intermediate Products, Final Products, and Waste Products Generated During Manufacture and Use:

- Hydrogen sulfide
- Sulfuric acid

### Citrus Processing

Raw Materials, Intermediate Products, Final Products, and Waste Products Generated During Manufacture and Use:

- Diphenyl

### Clay Pigeons

General Types of Associated Materials:

- Pitch
- Resins
- Solvents

## Cleaners

General Types of Associated Materials:

- Alcohols
- Solvents
- Surfactants

Raw Materials, Intermediate Products, Final Products, and Waste Product Generated During Manufacture and Use:

- n-Propyl alcohol

Other Associated Materials:

- Trichloroethene

## Cleaning Fluids

General Types of Associated Materials:

- Detergents
- Solvents
- Surfactants

Raw Materials, Intermediate Products, Final Products, and Waste Product Generated During Manufacture and Use:

- Acetone
- Ethylene glycol ether
- Methyl alcohol
- Tetrachloroethene
- Trichloroethane
- Trichloroethene
- Toluene
- Xylenes

### Cleansers, Wood

Raw Materials, Intermediate Products, Final Products, and Waste Products Generated During Manufacture and Use:

- Oxalic acid

### Cloth Manufacturing

General Types of Associated Materials:

- Acids
- Alkalis
- Detergents
- Dyes
- Flame retardants
- Fungicides
- Glues
- Soaps

Raw Materials, Intermediate Products, Final Products, and Waste Products Generated During Manufacture and Use:

- Aluminum salts
- Arsenic
- Formaldehyde
- Formaldehyde, dicyandiamide
- Potassium salts
- Sodium hydroxide
- Sodium metasilicate
- Sodium salts
- Sodium silicate
- Urea formaldehyde
- Zinc

Other Associated Materials:

- Calcium salts

## Clubs

• See Bars, Clubs, and Lounges

## Coal

General Types of Associated Materials:

• Acids
• Metals
• Solvents

Raw Materials, Intermediate Products, Final Products, and Waste Product Generated During Manufacture and Use:

• Carbon
• Creosote
• Lignins
• Phenols
• Polynuclear aromatic hydrocarbons
• Styrene

## Coal Gasification

General Types of Associated Materials:

• Coal tar

Raw Materials, Intermediate Products, Final Products, and Waste Product Generated During Manufacture and Use:

• Ammonia
• Chromium
• Cyanides
• Phenols
• Polynuclear aromatic hydrocarbons

Other Associated Materials:

- Benzene
- Xylenes

## Coal Oil, Synthetic

Raw Materials, Intermediate Products, Final Products, and Waste Products Generated During Manufacture and Use:

- Nickel

## Coal Tar and Distillates

General Types of Associated Materials:

- Coal
- Petroleum fuels
- Pitch
- Solvents

Raw Materials, Intermediate Products, Final Products, and Waste Products Generated During Manufacture and Use:

- Acetone
- Acridine
- Aniline
- Arsenic
- Creosote
- Cresol
- Lead
- Naphthalene
- Phenols
- Polynuclear aromatic hydrocarbons

Other Associated Materials:

- Benzene
- Benzol
- Ethyl benzene
- Toluene
- Xylenes

## Coffee, Decaffeinated

Raw Materials, Intermediate Products, Final Products, and Waste Product Generated During Manufacture and Use:

- Trichloroethene

## Coinage

General Types of Associated Materials:

- Lubricants
- Metals
- Solvents

Raw Materials, Intermediate Products, Final Products, and Waste Product Generated During Manufacture and Use:

- Aluminum
- Copper
- Nickel
- Silver
- Zinc

## Coke and Coke Ovens

General Types of Associated Materials:

- Coal tar and distillates

Raw Materials, Intermediate Products, Final Products, and Waste Products Generated During Manufacture and Use:

- Hydrogen cyanide
- Hydrogen sulfide
- Phenols
- Toluene
- Tricresyl phosphates

## Collodions

Raw Materials, Intermediate Products, Final Products, and Waste Products Generated During Manufacture and Use:

- Formates

Other Associated Materials:

- Ethyl ether
- Ketones

## Commutators

Raw Materials, Intermediate Products, Final Products, and Waste Products Generated During Manufacture and Use:

- Silver

### Compressors

General Types of Associated Materials:

- Lubricants
- Solvents

Other Associated Materials:

- Chloroform
- Methylene chloride
- Polychlorinated biphenyls
- Trichloroethane
- Trichloroethene

### Concrete

General Types of Associated Materials:

- Acids
- Portland cement

Raw Materials, Intermediate Products, Final Products, and Waste Product Generated During Manufacture and Use:

- Ketones
- Sodium

Other Associated Materials:

- Benzene
- Dimethoxane
- Dioxane
- Ethyl silicate

## Condensation Colors

Raw Materials, Intermediate Products, Final Products, and Waste Products Generated During Manufacture and Use:

- alpha-Naphthylamine

## Construction Sites

General Types of Associated Materials:

- Adhesives and removers
- Herbicides
- Lubricants
- Paints
- Petroleum fuels
- Pitch
- Sealers
- Solvents
- Wood preservatives

Raw Materials, Intermediate Products, Final Products, and Waste Products Generated During Manufacture and Use:

- Creosote
- Portland cement

## Copolymers

Other Associated Materials:

- Nitrosodimethylamine
- Styrene

## Copper Cleaning

General Types of Associated Materials:

- Acids
- Polishes

Raw Materials, Intermediate Products, Final Products, and Waste Produc
Generated During Manufacture and Use:

- Hydrogen fluoride

## Copper Etching

Raw Materials, Intermediate Products, Final Products, and Waste Produc
Generated During Manufacture and Use:

- Picric acid

## Copper Refining

Raw Materials, Intermediate Products, Final Products, and Waste Produc
Generated During Manufacture and Use:

- Arsenic
- Copper
- Creosote
- Fluorides
- Selenium
- Tellurium
- Xanthate

## Copper Sulfate

Raw Materials, Intermediate Products, Final Products, and Waste Products Generated During Manufacture and Use:

- Copper
- Sulfuric acid

## Cordite

Raw Materials, Intermediate Products, Final Products, and Waste Products Generated During Manufacture and Use:

- Ethylene glycol dinitrate
- Nitroglycerin

## Corrosion Inhibitors

Raw Materials, Intermediate Products, Final Products, and Waste Products Generated During Manufacture and Use:

- Ethanolamines
- Ethylenediamine

## Cosmetics

General Types of Associated Materials:

- Paraffin
- Waxes

Raw Materials, Intermediate Products, Final Products, and Waste Products Generated During Manufacture and Use:

- Bismuth
- Boron

Raw Materials, Intermediate Products, Final Products, and Waste Produc
Generated During Manufacture and Use (cont.):

- Dichloroethane
- Ethylene glycol ether
- Isopropyl alcohol
- Ketones
- n-Propyl alcohol
- Quinone
- Zinc

Other Associated Materials:

- Ethyl alcohol

## Cotton

Raw Materials, Intermediate Products, Final Products, and Waste Produc
Generated During Manufacture and Use:

- Chlorinated lime
- Ketones
- Potassium hydroxide
- Sodium hydroxide

## Cotton, Nitrated

General Types of Associated Materials:

- Cotton

Raw Materials, Intermediate Products, Final Products, and Waste Product
Generated During Manufacture and Use:

- Nitrates

Other Associated Materials:

- Furfural

## Coumaphos

General Types of Associated Materials:

- Organophosphorus insecticides

Raw Materials, Intermediate Products, Final Products, and Waste Products
Generated During Manufacture and Use:

- See Insecticides

## Crayons

General Types of Associated Materials:

- Metals
- Paraffins
- Waxes

Raw Materials, Intermediate Products, Final Products, and Waste Products
Generated During Manufacture and Use:

- Dichlorobenzidine

## Cream of Tartar

Raw Materials, Intermediate Products, Final Products, and Waste Products
Generated During Manufacture and Use:

- Oxalic acid

## Creosote

General Types of Associated Materials:

- Coal tar and distillates

Raw Materials, Intermediate Products, Final Products, and Waste Product Generated During Manufacture and Use:

- Benzene
- Cresol
- Guaiacol
- Phenols
- Polynuclear aromatic hydrocarbons
- Pyridine

## Cresote Oil

- See Creosote

## Crucibles

General Types of Associated Materials:

- Ceramics

Raw Materials, Intermediate Products, Final Products, and Waste Product Generated During Manufacture and Use:

- Graphite
- Platinum
- Zirconium

## Cumar

General Types of Associated Materials:

- Acetates

## Cutting Oils

Raw Materials, Intermediate Products, Final Products, and Waste Products Generated During Manufacture and Use:

- Ethanolamines
- Polychlorinated biphenyls
- Polynuclear aromatic hydrocarbons

## Cyanogen

Raw Materials, Intermediate Products, Final Products, and Waste Products Generated During Manufacture and Use:

- Hydrogen cyanide

## Cygon

General Types of Associated Materials:

- Organophosphate insecticides

Raw Materials, Intermediate Products, Final Products, and Waste Products Generated During Manufacture and Use:

- See Insecticides

### Dacron Fiber

Raw Materials, Intermediate Products, Final Products, and Waste Product Generated During Manufacture and Use:

- Phthalic anhydride

### Dairies

Raw Materials, Intermediate Products, Final Products, and Waste Product Generated During Manufacture and Use:

- Benzidine

### Dairy Farming

General Types of Associated Materials:

- Deodorants
- Detergents
- Pesticides
- Rodenticides

Raw Materials, Intermediate Products, Final Products, and Waste Product Generated During Manufacture and Use:

- Bacteria
- Viruses

Other Associated Materials:

- Polybrominated biphenyls

### Dalapon

General Types of Associated Materials:

- Chlorinated herbicides

Raw Materials, Intermediate Products, Final Products, and Waste Products Generated During Manufacture and Use:

- See Herbicides

### DDD

General Types of Associated Materials:

- Organochlorine insecticides

Raw Materials, Intermediate Products, Final Products, and Waste Products Generated During Manufacture and Use:

- See Insecticides

### DDE

General Types of Associated Materials:

- Organochlorine insecticides

Raw Materials, Intermediate Products, Final Products, and Waste Products Generated During Manufacture and Use:

- See Insecticides

## DDT

General Types of Associated Materials:

- Organochlorine insecticides

Raw Materials, Intermediate Products, Final Products, and Waste Produci
Generated During Manufacture and Use:

- See Insecticides

## DDVP

General Types of Associated Materials:

- Organophosphate insecticides

Raw Materials, Intermediate Products, Final Products, and Waste Produci
Generated During Manufacture and Use:

- See Insecticides

## Decalcifying

Raw Materials, Intermediate Products, Final Products, and Waste Produci
Generated During Manufacture and Use:

- Formic acid

## Defoliants

Raw Materials, Intermediate Products, Final Products, and Waste Produci
Generated During Manufacture and Use:

- Calcium cyanamide
- Dioxins

## Degreasers

General Types of Associated Materials:

- Alkalis
- Kerosene
- Ketones
- Solvents

Other Associated Materials:

- Acetone
- Carbon tetrachloride
- Chloroform
- Dichloroethyl ether
- Dioxane
- Methylene chloride
- Propylene dichloride
- Tetrachloroethene
- Toluene
- Trichloroethane
- Trichloroethene
- Xylenes

## Dehydrating Agents

Raw Materials, Intermediate Products, Final Products, and Waste Products
Generated During Manufacture and Use:

- Isopropyl alcohol
- Sulfuric acid

## Deicing compounds

Raw Materials, Intermediate Products, Final Products, and Waste Produc
Generated During Manufacture and Use:

- Isopropyl alcohol          .

## delta-BIIC

General Types of Associated Materials:

- Organochlorine insecticides

Raw Materials, Intermediate Products, Final Products, and Waste Produc
Generated During Manufacture and Use:

- See Insecticides

## Demeton-O

General Types of Associated Materials:

- Organophosphorus insecticides

Raw Materials, Intermediate Products, Final Products, and Waste Produc
Generated During Manufacture and Use:

- See Insecticides

## Demeton-S

General Types of Associated Materials:

- Organophosphorus insecticides

Raw Materials, Intermediate Products, Final Products, and Waste Products Generated During Manufacture and Use:

- See Insecticides

## Dental Alloys

General Types of Associated Materials:

- Amalgam

Raw Materials, Intermediate Products, Final Products, and Waste Products Generated During Manufacture and Use:

- Germanium
- Gold
- Mercury
- Platinum
- Silver

## Dental Cement

Raw Materials, Intermediate Products, Final Products, and Waste Products Generated During Manufacture and Use:

- Phosphoric acid

## Dentists

General Types of Associated Materials:

- Amalgams
- Anesthetics
- Dental alloys
- Disinfectants

General Types of Associated Materials (cont.):

- Resins
- Soaps

Raw Materials, Intermediate Products, Final Products, and Waste Products Generated During Manufacture and Use:

- Bacteria
- Eugenol
- Heavy metals
- Mercury

## Deodorants

Raw Materials, Intermediate Products, Final Products, and Waste Products Generated During Manufacture and Use:

- Chlorinated benzenes
- Formaldehyde
- Zinc
- Zirconium

Other Associated Materials:

- Dioxane

## Depilatories

Raw Materials, Intermediate Products, Final Products, and Waste Products Generated During Manufacture and Use:

- Hydrogen sulfide

## Desizing Agents

Raw Materials, Intermediate Products, Final Products, and Waste Products Generated During Manufacture and Use:

- n-Butylamine

## Detergent

General Types of Associated Materials:

- Naphtha

Raw Materials, Intermediate Products, Final Products, and Waste Products Generated During Manufacture and Use:

- Benzene
- Dioxane
- Ethanolamines
- Ethyl alcohol
- n-Butyl alcohol
- Phosphoric acid
- Sulfuric acid

## Detonators

Raw Materials, Intermediate Products, Final Products, and Waste Products Generated During Manufacture and Use:

- Tetryl

## Dewaxing

General Types of Associated Materials:

- Solvents

Raw Materials, Intermediate Products, Final Products, and Waste Produc
Generated During Manufacture and Use:

- n-Butylamine
- Ethylene chlorohydrin
- Ketones
- Methyl alcohol

## Dextrin

Raw Materials, Intermediate Products, Final Products, and Waste Produc
Generated During Manufacture and Use:

- Oxalic acid

## Diacetone

See Ketones

## Diaminoethylene

Raw Materials, Intermediate Products, Final Products, and Waste Produc
Generated During Manufacture and Use:

- Dichloroethane

## Diamonds, Synthetic

Raw Materials, Intermediate Products, Final Products, and Waste Produc
Generated During Manufacture and Use:

- Carbon
- Thallium

## Diazinon

General Types of Associated Materials:

- Organophosphorus insecticides

Raw Materials, Intermediate Products, Final Products, and Waste Products
Generated During Manufacture and Use:

- See Insecticides

## Dicamba

General Types of Associated Materials:

- Chlorinated herbicides

Raw Materials, Intermediate Products, Final Products, and Waste Products
Generated During Manufacture and Use:

- See Herbicides

## Dichloroprop

General Types of Associated Materials:

- Chlorinated herbicides

Raw Materials, Intermediate Products, Final Products, and Waste Products
Generated During Manufacture and Use:

- See Herbicides

## Dichlorvos

General Types of Associated Materials:

- Organophosphorus insecticides

Raw Materials, Intermediate Products, Final Products, and Waste Products
Generated During Manufacture and Use:

- See Insecticides

## Dicofol

General Types of Associated Materials:

- Organochlorine insecticides

Raw Materials, Intermediate Products, Final Products, and Waste Products
Generated During Manufacture and Use:

- See Insecticides

## Dieldrin

General Types of Associated Materials:

- Organochlorine insecticides

Raw Materials, Intermediate Products, Final Products, and Waste Products
Generated During Manufacture and Use:

- See Insecticides

## Diesel Fuel

Raw Materials, Intermediate Products, Final Products, and Waste Products Generated During Manufacture and Use:

- Benzene
- Creosote
- Ethyl benzene
- Polynuclear aromatic hydrocarbons
- Toluene
- Xylenes

## Diesel Fuel Additives

Raw Materials, Intermediate Products, Final Products, and Waste Products Generated During Manufacture and Use:

- Ethyl alcohol
- Ethyl ether
- Hydroquinone

## Dimethoate

General Types of Associated Materials:

- Organophosphate insecticides

Raw Materials, Intermediate Products, Final Products, and Waste Products Generated During Manufacture and Use:

- See Insecticides

## Dimethylhydrazine

Raw Materials, Intermediate Products, Final Products, and Waste Produc
Generated During Manufacture and Use:

- Nitrosodimethylamine

## Dinoseb

General Types of Associated Materials:

- Chlorinated herbicides

Raw Materials, Intermediate Products, Final Products, and Waste Produc
Generated During Manufacture and Use:

- See Herbicides

## Diolefins

Raw Materials, Intermediate Products, Final Products, and Waste Produc
Generated During Manufacture and Use:

- Dibromoethane

## Diphacin

General Types of Associated Materials:

- Rodenticides

Raw Materials, Intermediate Products, Final Products, and Waste Produc
Generated During Manufacture and Use:

- Coumarin
- Indandiones

## Diphenylamine

Raw Materials, Intermediate Products, Final Products, and Waste Products
Generated During Manufacture and Use:

- Aminodiphenyl

## Diquat

General Types of Associated Materials:

- Bipyridyl herbicides

Raw Materials, Intermediate Products, Final Products, and Waste Products
Generated During Manufacture and Use:

- See Herbicides

## Disinfectants

Raw Materials, Intermediate Products, Final Products, and Waste Products
Generated During Manufacture and Use:

- Acetaldehyde
- Acridine
- Aniline
- beta-Propiolactone
- Bismuth
- Chlorinated benzenes
- Chlorinated lime
- Chlorine
- Cresol
- Ethylene oxide
- Formaldehyde
- Furfural
- Hydrogen sulfide
- Iodine
- Isopropyl alcohol

Raw Materials, Intermediate Products, Final Products, and Waste Product Generated During Manufacture and Use (cont.):

- Manganese
- Mercury
- Mercury, alkyl
- Phenols
- Picric acid
- Propyl alcohol
- Sulfur dioxide
- Tetramethylthiuram disulfide
- Trichloroethene
- Zinc

Other Associated Materials:

- Ammonium compounds
- Carbolic acid
- Chlorine
- Pyridine

## Disinfectants, Foods

Raw Materials, Intermediate Products, Final Products, and Waste Product Generated During Manufacture and Use:

- Tetramethylthiuram disulfide

## Disulfoton

General Types of Associated Materials:

- Organophosphorus insecticides

Raw Materials, Intermediate Products, Final Products, and Waste Product Generated During Manufacture and Use:

- See Insecticides

## Dopes

General Types of Associated Materials:

- Acetates
- Ketones

Raw Materials, Intermediate Products, Final Products, and Waste Products Generated During Manufacture and Use:

- Boron hydrides

Other Associated Materials:

- n-Propyl alcohol
- Tetrachloroethene

## Dow Metal

Raw Materials, Intermediate Products, Final Products, and Waste Products Generated During Manufacture and Use:

- Magnesium

## Drill-glass

Raw Materials, Intermediate Products, Final Products, and Waste Products Generated During Manufacture and Use:

- Tin

## Drugs

Raw Materials, Intermediate Products, Final Products, and Waste Product
Generated During Manufacture and Use:

- Acridine
- Allyl alcohol
- Arsenic
- Benzyl chloride
- Chloroform
- Cobalt
- Dibromomethane
- Dichloroethene
- Dimethyl sulfate
- Ethyl chloride
- Ethyl ether
- Ethylene chlorohydrin
- Ethylenediamine
- Hexamethylenetetramine
- Hydrazine and derivatives
- Hydroquinone
- Manganese
- Mercury
- Molybdenum
- n-Butyl alcohol
- n-Butylamine
- Phenols
- Picric acid
- Platinum
- Propyl alcohol
- Silver
- Tetrachloroethane
- Trichloroethene

Other Associated Materials:

- Acetic anhydride
- Dimethylformamide
- Pyridine

**Dry Cleaning**

General Types of Associated Materials:

- Alkalis
- Bactericides
- Bleaches
- Brighteners
- Detergents
- Enzymes
- Fungicides
- Sizing
- Solvents
- Surfactants
- Turpentine
- Waterproofing

Other Associated Materials:

- Acetic acid
- Ammonia
- Amyl acetate
- Benzene
- Carbon disulfide
- Carbon tetrachloride
- Chlorine
- Dichloroethane
- Dichloroethyl ether
- Dichloroethene
- Ethyl ether
- Ethylene glycol ether
- Methanol
- Nitrobenzene
- Perchloroethene
- Polynuclear aromatic hydrocarbons
- Propylene dichloride
- Stoddard solvents
- Tetrachloroethane
- Tetrachloroethene
- Trichloroethane
- Trichloroethene

## Duplicating Fluids

General Types of Associated Materials:

- Alcohols
- Solvents

## Duprene

Raw Materials, Intermediate Products, Final Products, and Waste Products Generated During Manufacture and Use:

- Chloroprene

## Dursban

General Types of Associated Materials:

- Organophosphate insecticides

Raw Materials, Intermediate Products, Final Products, and Waste Products Generated During Manufacture and Use:

- See Insecticides

### Dyes

General Types of Associated Materials:

- Acetates
- Acids
- Alkalis
- Bleaches
- Coal tar products
- Detergents
- Dyes
- Gums
- Solvents

Raw Materials, Intermediate Products, Final Products, and Waste Products Generated During Manufacture and Use:

- Acetic acid
- Acetic anhydride
- Acridine
- alpha-Naphthylamine
- Aminodiphenyl
- Ammonia
- Aniline
- Antimony
- Arsenic
- Arsine
- Benzene
- Benzyl chloride
- beta-Naphthylamine
- Boron
- Bromine
- Carbon disulfide
- Chlorine
- Chlorodiphenyls
- Chromates
- Copper
- Cresol
- Cyanides
- Dextrins
- Dimethyl sulfate
- Dimethylaminoazobenzene

Raw Materials, Intermediate Products, Final Products, and Waste Product Generated During Manufacture and Use (cont.):

- Dinitro-o-cresol
- Dinitrobenzene
- Dinitrophenol
- Dinitrotoluene
- Ethanolamines
- Ethyl chloride
- Ethylenediamine
- Formaldehyde
- Formic acid
- Furfural
- Hydrogen fluoride
- Hydrogen sulfide
- Hydroquinones
- Iron
- Isopropyl alcohol
- Lead
- Manganese
- Mercaptans
- Mercury
- n-Butyl alcohol
- n-Butylamine
- Naphthalene
- Nickel
- Nitrobenzene
- Nitrogen
- Nitrophenol
- Oxalic acid
- Phenols
- Phosphates
- Phosphorus
- Phthalic anhydride
- Picric acid
- Potassium chlorate
- Pyridine
- Quinone
- Selenium
- Silicates
- Silver
- Sodium hydroxide

Raw Materials, Intermediate Products, Final Products, and Waste Products
Generated During Manufacture and Use (cont.):

- Sulfur chloride
- Sulfuric acid
- Thallium
- Tin
- Titanium
- Trichloroethene
- Vanadium
- Zinc

Other Associated Materials:

- Benzene
- Calcium salts
- Chlorinated benzenes
- Dimethylformamide
- Dioxane
- Ethyl alcohol
- Ethyl ether
- Ethylene chlorohydrin
- Ethylene glycol ether
- Hydrogen chloride
- Ketones
- Methyl alcohol
- Nitroparaffins
- Pyridine
- Xylenes

## Dyes, Alizarin

General Types of Associated Materials:

- Dyes

Raw Materials, Intermediate Products, Final Products, and Waste Products
Generated During Manufacture and Use:

- Phthalic anhydride

## Dyes, Aniline

General Types of Associated Materials:

- Aniline compounds
- Dyes

Raw Materials, Intermediate Products, Final Products, and Waste Product Generated During Manufacture and Use:

- Arsenic
- Nitrobenzene

## Dyes, Azo

General Types of Associated Materials:

- Dyes

Raw Materials, Intermediate Products, Final Products, and Waste Product Generated During Manufacture and Use:

- Benzidine

## Dynamite

Raw Materials, Intermediate Products, Final Products, and Waste Product Generated During Manufacture and Use:

- Ethylene glycol dinitrate
- Nitroglycerin

### Electric Condensers

Raw Materials, Intermediate Products, Final Products, and Waste Products Generated During Manufacture and Use:

- Chlorodiphenyls

### Edible Fat Processing

General Types of Associated Materials:

- Naphtha

Raw Materials, Intermediate Products, Final Products, and Waste Products Generated During Manufacture and Use:

- Acrolein
- Ethyl chloride
- Hydrogen sulfide
- Nickel
- Polynuclear aromatic hydrocarbons
- Sodium hydroxide
- Tetramethylthiuram disulfide

Other Associated Materials:

- Acetonitrile
- Amyl alcohol
- Benzene
- Carbon disulfide
- Carbon tetrachloride
- Cyclohexane
- Cyclohexene
- Cyclopropane
- Dibromomethane
- Dichloroethane
- Dichloroethyl ether
- Dichloroethene
- Dioxane

Other Associated Materials (cont.):

- Ethyl ether
- Formates
- Hydrogen chloride
- Ketones
- Methylcyclohexene
- Methylene chloride
- n-Hexane
- Nitroparaffins
- Propylene dichloride
- Tetrachloroethane
- Trichloroethene

## Edible Fats, Bacteriostats

Raw Materials, Intermediate Products, Final Products, and Waste Product:
Generated During Manufacture and Use:

- Tetramethylthiuram disulfide

## Edible Oil Processing

General Types of Associated Materials:

- Acetates
- Naphtha

Raw Materials, Intermediate Products, Final Products, and Waste Product:
Generated During Manufacture and Use:

- Acrolein
- Chlorinated lime
- Chromium
- Ethyl chloride
- Manganese
- Nickel

Raw Materials, Intermediate Products, Final Products, and Waste Products Generated During Manufacture and Use (cont.):

- Polynuclear aromatic hydrocarbons
- Sulfur chloride
- Sulfuric acid
- Tetramethylthiuram disulfide

Other Associated Materials:

- Acetonitrile
- Amyl alcohol
- Benzene
- Carbon disulfide
- Cyclohexane
- Cyclohexene
- Cyclopropane
- Dibromomethane
- Dichloroethane
- Dichloroethyl ether
- Dichloroethene
- Dioxane
- Ethyl ether
- Ethylene glycol ether
- Ethylenediamine
- Formates
- Hydrogen chloride
- Ketones
- Methylcyclohexene
- Methylene chloride
- n-Hexane
- Nitroparaffins
- Propylene dichloride
- Tetrachloroethane
- Trichloroethene

## Edible Oils, Bacteriostats

Raw Materials, Intermediate Products, Final Products, and Waste Product Generated During Manufacture and Use:

- Tetramethylthiuram disulfide

## EDTA

Raw Materials, Intermediate Products, Final Products, and Waste Product Generated During Manufacture and Use:

- Ethylenediamine

## Effluent Clarification

Raw Materials, Intermediate Products, Final Products, and Waste Product Generated During Manufacture and Use:

- Ethyleneimine

## Elastomers

Raw Materials, Intermediate Products, Final Products, and Waste Product Generated During Manufacture and Use:

- MOCA

## Electric Cables

Raw Materials, Intermediate Products, Final Products, and Waste Products Generated During Manufacture and Use:

- Chlorinated naphthalenes
- Polychlorinated biphenyls

## Electrical Components

Raw Materials, Intermediate Products, Final Products, and Waste Products Generated During Manufacture and Use:

- Boron
- Carbon disulfide
- Chlorinated naphthalenes
- Chlorodiphenyls
- Graphite
- Lead
- Mercury
- Molybdenum
- Nitrosodimethylamine
- Phthalates
- Platinum
- Polychlorinated biphenyls
- Thallium
- Tin

Other Associated Materials:

- Benzene
- Perchloroethene
- Trichloroethene

## Electrical Utilities

- See Utilities, Gas and Electric

## Electrodes

Raw Materials, Intermediate Products, Final Products, and Waste Product Generated During Manufacture and Use:

- Graphite
- Selenium
- Titanium

## Electroluminescent Coatings

Raw Materials, Intermediate Products, Final Products, and Waste Product Generated During Manufacture and Use:

- Phosphorus

## Electroplating

General Types of Associated Materials:

- Abrasives
- Acids
- Alkalis
- Chlorinated waxes
- Degreasers
- Detergents
- Gasoline
- Lubricants
- Metal cleaners
- Naphtha
- Petroleum fuels
- Soaps
- Soluble oils
- Solvents
- Waste oils

Raw Materials, Intermediate Products, Final Products, and Waste Products Generated During Manufacture and Use:

- Aluminum
- Ammonia
- Arsenic
- Arsine
- Asbestos
- Boron
- Chlorides
- Chromic acid
- Chromium
- Cobalt
- Copper
- Cyanides
- Dichloroethane
- Formic acid
- Germanium
- Heavy metals
- Hydrochloric acid
- Hydrogen cyanide
- Iron
- Lead
- Mercury
- Molybdenum
- Nickel
- Nitrates
- Nitrogen
- Phosphoric acid
- Platinum
- Potassium hydroxide
- Silver
- Sulfate
- Sulfuric acid
- Triethanolamine
- Zinc

Other Associated Materials:

- Ammonia
- Benzene
- Carbon disulfide
- Chloroform
- Chromic acid
- Dichloroethene
- Ethyl benzene
- Hexane
- Hydrogen chloride
- Perchloroethene
- Polychlorinated biphenyls
- Tetrachloroethene
- Toluene
- Trichloroethane
- Trichloroethene
- Xylenes

## Elemi

General Types of Associated Materials:

- Acetates

## Embalming

General Types of Associated Materials:

- Acids
- Alkalis
- Oil of cinnamon
- Oil of clove

Raw Materials, Intermediate Products, Final Products, and Waste Products Generated During Manufacture and Use:

- Antimony
- Arsenic
- Bacteria
- Barium
- Cadmium
- Chromium
- Cobalt
- Formaldehyde
- Lead
- Mercury
- Methyl alcohol
- Nickel
- Phenols
- Thymol
- Zinc

## Emulsifying Agents

Raw Materials, Intermediate Products, Final Products, and Waste Products Generated During Manufacture and Use:

- Dioxane
- Ethanolamines
- Ethyl benzene
- Ethylenediamine
- n-Butylamine

## Enamels

General Types of Associated Materials:

- Pigments
- Solvents

Raw Materials, Intermediate Products, Final Products, and Waste Product
Generated During Manufacture and Use:

- Antimony
- Arsenic
- Boron
- Cerium
- Fluorides
- Lead
- Nickel
- Tin
- Uranium
- Zirconium

Other Associated Materials:

- Methyl alcohol

## Endosulfan I

General Types of Associated Materials:

- Organochlorine insecticides

Raw Materials, Intermediate Products, Final Products, and Waste Product
Generated During Manufacture and Use:

- See Insecticides

## Endosulfan II

General Types of Associated Materials:

- Organochlorine insecticides

Raw Materials, Intermediate Products, Final Products, and Waste Products Generated During Manufacture and Use:

- See Insecticides

## Endosulfan Sulfate

General Types of Associated Materials:

- Organochlorine insecticides

Raw Materials, Intermediate Products, Final Products, and Waste Products Generated During Manufacture and Use:

- See Insecticides

## Engraving

General Types of Associated Materials:

- Abrasives
- Acids
- Alkalis
- Solvents

Raw Materials, Intermediate Products, Final Products, and Waste Products Generated During Manufacture and Use:

- Cadmium
- Chromium
- Cyanides
- Ferric chloride
- Phosphoric acid

Other Associated Materials:

- Chromic acid

## EPN

General Types of Associated Materials:

- Organophosphate insecticides

Raw Materials, Intermediate Products, Final Products, and Waste Products Generated During Manufacture and Use:

- See Insecticides

## Epoxy Resins

General Types of Associated Materials:

- Resins
- Solvents

Raw Materials, Intermediate Products, Final Products, and Waste Products Generated During Manufacture and Use:

- Epichlorohydrin
- Ketones
- MOCA
- Styrene
- Xylenes

## Erythrosin

Raw Materials, Intermediate Products, Final Products, and Waste Products Generated During Manufacture and Use:

- Phthalic anhydride

## Essences

Raw Materials, Intermediate Products, Final Products, and Waste Products Generated During Manufacture and Use:

- Formates

## Esters

Raw Materials, Intermediate Products, Final Products, and Waste Products Generated During Manufacture and Use:

- Acetic acid

Other Associated Materials:

- Methyl alcohol
- Methylene chloride
- Nitroparaffins
- Tetrachloroethene

## Esters, Acrylic

General Types of Associated Materials:

- Esters

Raw Materials, Intermediate Products, Final Products, and Waste Products Generated During Manufacture and Use:

- Nickel

## Esters, Methyl

General Types of Associated Materials:

- Esters

Raw Materials, Intermediate Products, Final Products, and Waste Produc
Generated During Manufacture and Use:

- Dimethyl sulfate

## Esters, Polymethacrylic

General Types of Associated Materials:

- Esters

Raw Materials, Intermediate Products, Final Products, and Waste Produc
Generated During Manufacture and Use:

- Tricresyl phosphates

## Etching

General Types of Associated Materials:

- Acids
- Alkalis

Raw Materials, Intermediate Products, Final Products, and Waste Produc
Generated During Manufacture and Use:

- Aluminum salts
- Antimony
- Arsenic
- Arsine
- Chromates

Raw Materials, Intermediate Products, Final Products, and Waste Products Generated During Manufacture and Use (cont.):

- Copper
- Iron
- Lead
- Nitrogen
- Phosphates
- Phthalates
- Picric acid
- Potassium hydroxide
- Silicates
- Silver
- Stibine
- Tin
- Zinc

## Ethanolamine

Raw Materials, Intermediate Products, Final Products, and Waste Products Generated During Manufacture and Use:

- Ethylene oxide

## Ethers

Raw Materials, Intermediate Products, Final Products, and Waste Products Generated During Manufacture and Use:

- Dimethyl sulfate

Other Associated Materials:

- Tetrachloroethene

## Enthion

General Types of Associated Materials:

- Organophosphate insecticides

Raw Materials, Intermediate Products, Final Products, and Waste Products Generated During Manufacture and Use:

- See Insecticides

## Ethoprop

General Types of Associated Materials:

- Organophosphorus insecticides

Raw Materials, Intermediate Products, Final Products, and Waste Products Generated During Manufacture and Use:

- See Insecticides

## Ethyl Alcohol

Raw Materials, Intermediate Products, Final Products, and Waste Products Generated During Manufacture and Use:

- Acetic acid
- Amyl alcohol
- Methyl alcohol

### Ethyl Alcohol, Denaturers

Raw Materials, Intermediate Products, Final Products, and Waste Products
Generated During Manufacture and Use:

- Pyridine

### Ethyl Cellulose

General Types of Associated Materials:

- Acetates
- Cellulose

Raw Materials, Intermediate Products, Final Products, and Waste Products
Generated During Manufacture and Use:

- Dichloroethyl ether
- Ethyl chloride

Other Associated Materials:

- Ethylene chlorohydrin
- n-Propyl alcohol

### Ethyl Chloride

Raw Materials, Intermediate Products, Final Products, and Waste Products
Generated During Manufacture and Use:

- Ethyl alcohol

## Ethyl Ether

Raw Materials, Intermediate Products, Final Products, and Waste Product Generated During Manufacture and Use:

- Ethyl alcohol

## Ethyl Glycol

Raw Materials, Intermediate Products, Final Products, and Waste Product Generated During Manufacture and Use:

- Dichloroethane

## Ethylating Agents

Raw Materials, Intermediate Products, Final Products, and Waste Product Generated During Manufacture and Use:

- Ethyl chloride

## Ethylene Glycol

Raw Materials, Intermediate Products, Final Products, and Waste Product Generated During Manufacture and Use:

- Ethylene chlorohydrin
- Ethylene oxide
- Methyl alcohol

## Ethylene Oxide

Raw Materials, Intermediate Products, Final Products, and Waste Products Generated During Manufacture and Use:

- Ethylene chlorohydrin

## Explosives

Raw Materials, Intermediate Products, Final Products, and Waste Products Generated During Manufacture and Use:

- Acetaldehyde
- Amyl alcohol
- Antimony
- Benzene
- Carbon disulfide
- Dinitrobenzene
- Dinitrophenol
- Dinitrotoluene
- Ethyl ether
- Ethylene glycol
- Ethylene glycol dinitrate
- Formaldehyde
- Graphite
- Hexamethylenetetramine
- Ketones
- Mercury
- Nitrobenzene
- Nitroglycerin
- PETN
- Phenols
- Phosphorus
- Picric acid
- Pyridine
- Sodium hydroxide
- Sulfuric acid
- Tetryl
- Trinitrotoluene
- Zirconium

Other Associated Materials:

- Acetic anhydride
- Ammonium compounds
- Ethyl alcohol

## Exterminators

General Types of Associated Materials:

- Insecticides
- Herbicides
- Molluscicides
- Pesticides
- Rodenticides

## Extraction Agents

Raw Materials, Intermediate Products, Final Products, and Waste Products Generated During Manufacture and Use:

- Acetone
- Isopropyl alcohol
- Methylene chloride
- Tetrachloroethene
- Trichloroethane
- Trichloroethene

## Face Creams

Raw Materials, Intermediate Products, Final Products, and Waste Products Generated During Manufacture and Use:

- Ethanolamines

## Farming

General Types of Associated Materials:

- Detergents
- Disinfectants
- Feeds, treated
- Fertilizers
- Fungicides
- Herbicides
- Kerosene
- Lubricants
- Oils
- Paints
- Pesticides
- Petroleum hydrocarbons
- Seed treatment
- Solvents
- Tar and derivatives
- Wood preservatives

Raw Materials, Intermediate Products, Final Products, and Waste Products Generated During Manufacture and Use:

- Bacteria
- Cadmium
- Carbon tetrachloride
- Copper
- Dibromomethane
- Ethanolamines
- Ethylene oxide
- Formates
- Heavy metals
- Lead
- Manure
- Mercury
- Nitrates
- Polybrominated biphenyls
- Polychlorinated biphenyls
- Tetrachloroethane
- Trichloroethene
- Zinc

## Farming, Mushrooms

General Types of Associated Materials:

- Farming

Raw Materials, Intermediate Products, Final Products, and Waste Product: Generated During Manufacture and Use:

- Tetramethylthiuram disulfide

## Farming, Potatoes

General Types of Associated Materials:

- Farming

Raw Materials, Intermediate Products, Final Products, and Waste Product: Generated During Manufacture and Use:

- Amyl alcohol

Other Associated Materials:

- Ethylene chlorohydrin

## Feathers

General Types of Associated Materials:

- Detergents
- Solvents

Other Associated Materials:

- Methyl alcohol

## Felt Processing

General Types of Associated Materials:

- Acids
- Dyes
- Metals

Raw Materials, Intermediate Products, Final Products, and Waste Products Generated During Manufacture and Use:

- Anthrax bacillus
- Bacteria
- Glauber's salt
- Hydrogen peroxide
- Hydrogen sulfide
- Mercury
- Quinones
- Sodium carbonate

Other Associated Materials:

- Methyl alcohol

## Fensulfothion

General Types of Associated Materials:

- Organophosphorus insecticides

Raw Materials, Intermediate Products, Final Products, and Waste Products Generated During Manufacture and Use:

- See Insecticides

## Fenthion

General Types of Associated Materials:

• Organophosphate insecticides

Raw Materials, Intermediate Products, Final Products, and Waste Products Generated During Manufacture and Use:

• See Insecticides

## Ferbam

General Types of Associated Materials:

• Dithiocarbamate fungicides

Raw Materials, Intermediate Products, Final Products, and Waste Products Generated During Manufacture and Use:

• See Fungicides

## Fermentation Inhibitors

Raw Materials, Intermediate Products, Final Products, and Waste Products Generated During Manufacture and Use:

• Fluorides

## Ferrovanadium Products

Raw Materials, Intermediate Products, Final Products, and Waste Products Generated During Manufacture and Use:

• Iron
• Vanadium

## Fertilizer

General Types of Associated Materials:

- Acids
- Herbicides
- Pesticides

Raw Materials, Intermediate Products, Final Products, and Waste Products Generated During Manufacture and Use:

- Ammonia
- Arsine
- Calcium cyanamide
- Fluorides
- Hydrogen sulfide
- Manganese
- Manure
- Molybdenum
- Nitrates
- Nitrogen
- Phenols
- Phosphates
- Phosphoric acid
- Phosphorus
- Potassium salts
- Sulfuric acid

Other Associated Materials:

- Ammonium compounds
- Hydrogen chloride

## Fiber Glass

General Types of Associated Materials:

- Solvents

Raw Materials, Intermediate Products, Final Products, and Waste Products Generated During Manufacture and Use:

- Resins
- Styrene

## Films

General Types of Associated Materials:

- Cellulose
- Plastics

Raw Materials, Intermediate Products, Final Products, and Waste Products Generated During Manufacture and Use:

- Benzidine

Other Associated Materials:

- Ethylene glycol ether

## Fire Extinguishers

Raw Materials, Intermediate Products, Final Products, and Waste Products Generated During Manufacture and Use:

- Chloroform
- Carbon tetrachloride
- Dibromomethane
- Sodium bicarbonate

## Fire Retardants

Raw Materials, Intermediate Products, Final Products, and Waste Products Generated During Manufacture and Use:

- Tricresyl phosphates

## Fire Stations and Sites

General Types of Associated Materials:

- Alcohols
- Lubricants
- Petroleum fuels
- Solvents

Raw Materials, Intermediate Products, Final Products, and Waste Products Generated During Manufacture and Use:

- Carbon tetrachloride
- Chloroform
- Dibromomethane
- Dioxin
- Freon 11
- Freon 11C
- Ketones
- Sodium bicarbonate
- Tetrachloroethene
- Tetrahydrofurans
- Trichlorobenzene
- Trichloroethene

## Fireworks

General Types of Associated Materials:

- Explosives
- Incendiaries

Raw Materials, Intermediate Products, Final Products, and Waste Product Generated During Manufacture and Use:

- Aluminum
- Antimony
- Arsenic
- Manganese
- Picric acid
- Thallium

## Fish Processing

- See Meat Packaging

## Flameproofing

Raw Materials, Intermediate Products, Final Products, and Waste Product Generated During Manufacture and Use:

- Antimony
- Boron
- Chlorodiphenyls
- Ethyleneimine

## Flares

General Types of Associated Materials:

- Explosives
- Incendiaries

Raw Materials, Intermediate Products, Final Products, and Waste Product Generated During Manufacture and Use:

- Magnesium

## Flavorings

Raw Materials, Intermediate Products, Final Products, and Waste Products Generated During Manufacture and Use:

- Acetaldehyde
- Formates

Other Associated Materials:

- Acetic anhydride
- Methylene chloride

## Flocculation Products

Raw Materials, Intermediate Products, Final Products, and Waste Products Generated During Manufacture and Use:

- Ethyleneimine

## Florists

General Types of Associated Materials:

- Fertilizers
- Fungicides
- Herbicides
- Insecticides

Raw Materials, Intermediate Products, Final Products, and Waste Products Generated During Manufacture and Use:

- Bacteria
- Heavy metals

## Flotation Agents

Raw Materials, Intermediate Products, Final Products, and Waste Product Generated During Manufacture and Use:

- Carbon disulfide
- Phosphorus

## Flour Mills

General Types of Associated Materials:

- Bleaches
- Pesticides
- Rodenticides

## Fluorescent Lamps

Raw Materials, Intermediate Products, Final Products, and Waste Product Generated During Manufacture and Use:

- Cadmium

## Fluorination

Raw Materials, Intermediate Products, Final Products, and Waste Product Generated During Manufacture and Use:

- Hydrogen fluoride

## Fluorocarbons

Raw Materials, Intermediate Products, Final Products, and Waste Products
Generated During Manufacture and Use:

- Carbon tetrachloride
- Chloroform
- Freon

## Foil

Raw Materials, Intermediate Products, Final Products, and Waste Products
Generated During Manufacture and Use:

- Aluminum
- Tin

## Food Processing

General Types of Associated Materials:

- Bleaches
- Brine
- Dyes
- Enzymes
- Petroleum fuels
- Resins
- Salts
- Waxes

Raw Materials, Intermediate Products, Final Products, and Waste Products
Generated During Manufacture and Use:

- Acetaldehyde
- Bacteria
- Chlorides
- Dichloroethane

Raw Materials, Intermediate Products, Final Products, and Waste Products Generated During Manufacture and Use (cont.):

- Iron, ferrous
- Lime
- Sulfuric acid

Other Associated Materials:

- Hydrogen chloride

## Foot Wear

General Types of Associated Materials:

- Acetates
- Dyes
- Glues
- Leather
- Plastics

## Formaldehyde

Raw Materials, Intermediate Products, Final Products, and Waste Products Generated During Manufacture and Use:

- Formalin
- Methyl alcohol

## Foundries

General Types of Associated Materials:

- Acids
- Glues
- Metals
- Petroleum fuels
- Resins
- Solvents

Raw Materials, Intermediate Products, Final Products, and Waste Products Generated During Manufacture and Use:

- Aluminum
- Barium
- Boron
- Calcium cyanamide
- Chlorinated naphthalenes
- Chlorodiphenyls
- Chromium
- Cresol
- Graphite
- Heavy metals
- Hydrogen fluoride
- Lead
- Magnesium oxide
- Manganese
- Nickel
- Phenols
- Polychlorinated biphenyls
- Tellurium
- Zirconium

Other Associated Materials:

- Benzene
- Dichloroethane
- Ethyl benzene
- Isopropanol
- Methanol

Other Associated Materials (cont.):

- Methyl alcohol
- Methylene chloride
- Toluene
- Trichloroethane
- Xylenes

## Fuel Oil

Raw Materials, Intermediate Products, Final Products, and Waste Products Generated During Manufacture and Use:

- Benzene
- Creosote
- Ethyl benzene
- Polynuclear aromatic hydrocarbons
- Toluene
- Xylenes

## Fuels, Solid

Raw Materials, Intermediate Products, Final Products, and Waste Products Generated During Manufacture and Use:

- Hexamethylenetetramine

## Fumarin

General Types of Associated Materials:

- Rodenticides

Raw Materials, Intermediate Products, Final Products, and Waste Products Generated During Manufacture and Use:

- Coumarin
- Indandiones

## Fumigants

Raw Materials, Intermediate Products, Final Products, and Waste Products Generated During Manufacture and Use:

- Acrylonitrile
- Boron
- Carbon disulfide
- Carbon tetrachloride
- Chlorinated benzenes
- Dibromomethane
- Dichloroethane
- Dichloroethyl ether
- Dioxane
- Ethylene dibromide
- Ethylene oxide
- Formaldehyde
- Formates
- Hydrogen cyanide
- Mercaptans
- Methyl bromide
- Methyl formate
- Methylene chloride
- Naphthalene
- Perchloroethene
- Propylene dichloride
- Stibine

Raw Materials, Intermediate Products, Final Products, and Waste Products Generated During Manufacture and Use (cont.):

- Sulfur dioxide
- Tetrachloroethane
- Tetrachloroethene
- Trichloroethene

## Fumigants, Grain

Raw Materials, Intermediate Products, Final Products, and Waste Products Generated During Manufacture and Use:

- Phosphides
- Carbon disulfide

## Fungicides

Raw Materials, Intermediate Products, Final Products, and Waste Products Generated During Manufacture and Use:

- Allyl alcohol
- Boron
- Cadmium
- Calcium oxide
- Chromium
- Copper
- Cyclohexane
- Cyclohexene
- Cyclopropane
- Diphenyl
- Ethylene oxide
- Ethylenediamine
- Formaldehyde
- Furfural
- Magnesium
- Mercaptans

Raw Materials, Intermediate Products, Final Products, and Waste Products
Generated During Manufacture and Use (cont.):

- Mercury
- Mercury, alkyl
- Methylcyclohexene
- Naphthalene
- Pentachlorophenol
- Phenols
- Tetramethylthiuram disulfide
- Thallium
- Tin
- Zinc

## Fungistats

Raw Materials, Intermediate Products, Final Products, and Waste Products
Generated During Manufacture and Use:

- Diphenyl

## Fur Processing

General Types of Associated Materials:

- Acids
- Alkalis
- Bleaches
- Dyes
- Oils
- Salts

Raw Materials, Intermediate Products, Final Products, and Waste Products
Generated During Manufacture and Use:

- Alum
- Bacteria

Raw Materials, Intermediate Products, Final Products, and Waste Products Generated During Manufacture and Use (cont.):

- Chromates
- Formaldehyde
- Hydroquinone
- Sulfuric acid
- Tetrachloroethane

## Furnaces

Raw Materials, Intermediate Products, Final Products, and Waste Products Generated During Manufacture and Use:

- Sulfur dioxide

## Furniture Manufacturing

General Types of Associated Materials:

- Acids
- Alkalis
- Naphtha
- Paints
- Polishes
- Soaps
- Solvents
- Stains
- Turpentine
- Waxes

Raw Materials, Intermediate Products, Final Products, and Waste Products Generated During Manufacture and Use:

- Cadmium
- Chromium
- Formaldehyde
- Lead

Raw Materials, Intermediate Products, Final Products, and Waste Products Generated During Manufacture and Use (cont.):

- Pyridine
- Rosin
- Sodium carbamate
- Sodium gravnomate
- Sodium hydroxide
- Zinc

Other Associated Materials:

- Acetone
- Benzene
- Ketones
- Methyl alcohol
- Methylene chloride
- Perchloroethene
- Toluene
- Xylenes

## Fuses

Raw Materials, Intermediate Products, Final Products, and Waste Products Generated During Manufacture and Use:

- Bismuth

## Galvanizing

General Types of Associated Materials and Processes

- Plating and polishing

Raw Materials, Intermediate Products, Final Products, and Waste Product Generated During Manufacture and Use:

- Ammonia
- Arsine
- Zinc

Other Associated Materials:

- Hydrogen chloride

# Gardening

- See Farming, Florists

# Gardona

General Types of Associated Materials:

- Organophosphate insecticides

Raw Materials, Intermediate Products, Final Products, and Waste Product Generated During Manufacture and Use:

- See Insecticides

# Gas Absorbents

Raw Materials, Intermediate Products, Final Products, and Waste Product Generated During Manufacture and Use:

- Ethanolamines

## Gas Mantles

Raw Materials, Intermediate Products, Final Products, and Waste Products
Generated During Manufacture and Use:

- Beryllium
- Thorium

## Gas Odorants

Raw Materials, Intermediate Products, Final Products, and Waste Products
Generated During Manufacture and Use:

- Mercaptans

## Gas Processing

Raw Materials, Intermediate Products, Final Products, and Waste Products
Generated During Manufacture and Use:

- Benzene
- Mercaptans
- Phenols

## Gas Service Station

- See Automotive Service Stations

## Gas Station

See Automotive Service Stations

## Gas Utilities

- See Utilities, Gas and Electric

## Gas Works

Raw Materials, Intermediate Products, Final Products, and Waste Products Generated During Manufacture and Use:

- Benzene
- Hydrogen cyanide
- Hydrogen sulfide

## Gaskets

General Types of Associated Materials:

- Cork
- Leather
- Plastics
- Rubber
- Solvents

## Gasoline

Raw Materials, Intermediate Products, Final Products, and Waste Products Generated During Manufacture and Use:

- Benzene
- Ethyl benzene
- Toluene
- Xylenes

### Gasoline Additives

General Types of Associated Materials:

- Alcohols
- Gasoline

Raw Materials, Intermediate Products, Final Products, and Waste Products Generated During Manufacture and Use:

- Benzyl chloride
- Boron hydrides
- Dibromoethane
- Dibromomethane
- Dichloroethane
- Ethyl alcohol
- Ethyl ether
- Ethylene oxide
- Hydroquinone
- Lead
- Lead, alkyl
- Methyl tertiary butyl ether
- Phosphorus
- Platinum
- Tricresyl phosphates

### Gasoline Station

- See Automotive Service Stations

### Gelatin

Raw Materials, Intermediate Products, Final Products, and Waste Products Generated During Manufacture and Use:

- Quinone
- Phosphoric acid
- Sulfur dioxide

## Gems, Synthetic

Raw Materials, Intermediate Products, Final Products, and Waste Product,
Generated During Manufacture and Use:

- Beryllium
- Boron

## Germicides

Raw Materials, Intermediate Products, Final Products, and Waste Product:
Generated During Manufacture and Use:

- Benzyl chloride
- Formaldehyde
- Furfural
- Mercury

## Glass Working

General Types of Associated Materials:

- Acetates
- Cleaners
- Resins
- Solvents
- Turpentine

Raw Materials, Intermediate Products, Final Products, and Waste Product:
Generated During Manufacture and Use:

- Antimony
- Arsenic
- Boron
- Calcium oxide
- Carbon disulfide
- Cerium

Raw Materials, Intermediate Products, Final Products, and Waste Products Generated During Manufacture and Use (cont.):

- Chromium
- Cobalt
- Fluorides
- Germanium
- Heavy metals
- Hydrofluoric acid
- Hydrogen fluoride
- Lead
- Manganese
- Methyl alcohol
- Molybdenum
- Nickel
- Nitrogen
- Phthalates
- Picric acid
- Selenium
- Silver
- Soda ash
- Sulfur dioxide
- Tellurium
- Tin
- Uranium
- Vanadium
- Zirconium

Other Associated Materials:

- Borax
- Boric acid
- Isopropyl alcohol
- Trichloroethene

## Glazing

General Types of Associated Materials:

- Ceramics
- Glass
- Metals

Raw Materials, Intermediate Products, Final Products, and Waste Products Generated During Manufacture and Use:

- Antimony
- Boron
- Cadmium
- Lead
- Tellurium
- Uranium
- Vanadium

## Glove Manufacturing

General Types of Associated Materials:

- Cotton
- Dyes
- Leather
- Plastics
- Rubber

Raw Materials, Intermediate Products, Final Products, and Waste Products Generated During Manufacture and Use:

- Phthalates
- Polyvinyl chloride
- Vinyl chloride

## Glucose

Raw Materials, Intermediate Products, Final Products, and Waste Products Generated During Manufacture and Use:

- Sulfuric acid

## Glues

Raw Materials, Intermediate Products, Final Products, and Waste Products Generated During Manufacture and Use:

- Ammonia
- Ethylene glycol
- Sulfuric acid

Other Associated Materials:

- Bacteria
- Benzene
- Dioxane
- Hydrogen chloride

## Glycerine

Raw Materials, Intermediate Products, Final Products, and Waste Products Generated During Manufacture and Use:

- Allyl alcohol
- Oxalic acid

## Glycerol

Raw Materials, Intermediate Products, Final Products, and Waste Products Generated During Manufacture and Use:

Epichlorohydrin
Oxalic acid

## Glycerophosphoric Acid

Raw Materials, Intermediate Products, Final Products, and Waste Products Generated During Manufacture and Use:

- Epichlorohydrin
- Glycerol
- Phosphoric Acid

## Glycidol Derivatives

Raw Materials, Intermediate Products, Final Products, and Waste Products Generated During Manufacture and Use:

- Epichlorohydrin

## Gold Extraction

Raw Materials, Intermediate Products, Final Products, and Waste Products Generated During Manufacture and Use:

- Arsenic
- Bromine
- Heavy metals
- Hydrogen cyanide
- Mercury
- Sulfur chloride

Other Associated Materials:

- Fluorides


## Grain Fermentation

Raw Materials, Intermediate Products, Final Products, and Waste Products Generated During Manufacture and Use:

- Amyl alcohol


## Grains Storage and Treating

General Types of Associated Materials and Activities

- Farming
- Fertilizers
- Heavy metals
- Pesticides

Raw Materials, Intermediate Products, Final Products, and Waste Products Generated During Manufacture and Use:

- Ammonium nitrate
- Carbon tetrachloride
- Chlorinated benzenes
- Ethylene oxide
- Formates
- Mercury
- Nitrogen
- Phosphides
- Urea

Other Associated Materials:

- Carbon disulfide

## Graphite

Raw Materials, Intermediate Products, Final Products, and Waste Products Generated During Manufacture and Use:

- Carbon
- Fluorides

## Grease

- See Lubricants

## Greenhouses

- See Florists

## Grinding

- See Abrasives

## Gums

Other Associated Materials:

- Dibromomethane
- Dichloroethyl ether
- Epichlorohydrin
- Ethyl ether
- Furfural
- Ketones
- n-Propyl alcohol
- Propylene dichloride
- Trichloroethene
- Tetrachloroethene

## Hair Dressers

• See Barber and Beauty Shops

## Hair Dyes

• See Dyes

## Hair Wave Preparations

General Types of Associated Materials:

- Alcohols
- Caustics

## Hand Creams

Raw Materials, Intermediate Products, Final Products, and Waste Products Generated During Manufacture and Use:

- Ethanolamines

## Hats

• See Felt Processing

## Hats, Straw

General Types of Associated Materials:

- Bleaches
- Dyes
- Glues

## Heptachlor

General Types of Associated Materials:

- Organochlorine insecticides

Raw Materials, Intermediate Products, Final Products, and Waste Product. Generated During Manufacture and Use:

- See Insecticides

## Heptachlor Epoxide

General Types of Associated Materials:

- Organochlorine insecticides

Raw Materials, Intermediate Products, Final Products, and Waste Product: Generated During Manufacture and Use:

- See Insecticides

## Herbicides

General Types of Associated Materials:

- Kerosene
- Petroleum oils

Raw Materials, Intermediate Products, Final Products, and Waste Products Generated During Manufacture and Use:

- Allyl alcohol
- Amino triazole
- Ammonium sulfamate
- Arsenic
- Arsenic trioxide
- Calcium cyanamide
- Carbamate derivatives
- Chlorodiphenyls
- Copper
- Copper sulfate
- Crag herbicide
- Creosote
- Cresol
- Dinitro-o-cresol
- Dinitrophenol
- Dinitrophenols
- Ethanolamines
- Furfural
- Mercury
- Pentachlorophenol
- Phenols
- Phenoxyacetic acid derivatives
- Phenylmercuric acetate
- Sodium arsenate
- Sodium borate
- Sodium chlorate
- Tetrachloroethane
- Trichloroacetic acid

## Hide Preservation

- See Tanneries

## High Density Liquids

General Types of Associated Materials:

- Brines

Raw Materials, Intermediate Products, Final Products, and Waste Product Generated During Manufacture and Use:

- Thallium

## Highway Maintenance

General Types of Associated Materials:

- Asphalt
- Brine
- Lubricants
- Salts
- Solvents
- Tar

Raw Materials, Intermediate Products, Final Products, and Waste Product Generated During Manufacture and Use:

- Chlorides
- Heavy metals
- Polychlorinated biphenyls
- Polynuclear aromatic hydrocarbons

Other Associated Materials:

- Benzene
- Ethyl benzene
- Methylene chloride
- Tetrachloroethene
- Toluene
- Trichloroethane
- Trichloroethene
- Xylenes

## Hoists

General Types of Associated Materials:

- Diesel fuel
- Hydraulic fluids

Raw Materials, Intermediate Products, Final Products, and Waste Products Generated During Manufacture and Use:

- Polychlorinated biphenyls
- Polynuclear aromatic hydrocarbons

## Horticulture

- See Farming

## Hydraulic Fluids

General Types of Associated Materials:

- Petroleum hydrocarbons

Raw Materials, Intermediate Products, Final Products, and Waste Products Generated During Manufacture and Use:

- Amyl alcohol
- Ethylene glycol
- Ethylene glycol ether
- Ketones
- Phosphorus
- Polychlorinated biphenyls
- Polynuclear aromatic hydrocarbons
- Tricresyl phosphates

## Hydrocarbon Bromination

Raw Materials, Intermediate Products, Final Products, and Waste Products
Generated During Manufacture and Use:

- Bromine
- Petroleum hydrocarbons

## Hydrochloric Acid

Raw Materials, Intermediate Products, Final Products, and Waste Products
Generated During Manufacture and Use:

- Sulfuric acid

Other Associated Materials:

- Hydrogen chloride

## Hydrocyanic Acid

Raw Materials, Intermediate Products, Final Products, and Waste Products
Generated During Manufacture and Use:

- Hydrogen cyanide

## Hydrofluoric Acid

Raw Materials, Intermediate Products, Final Products, and Waste Products
Generated During Manufacture and Use:

- Hydrogen fluoride

### Hydrogen Cyanide

Raw Materials, Intermediate Products, Final Products, and Waste Products
Generated During Manufacture and Use:

- Oxalic acid

### Hydrogen Peroxide

Raw Materials, Intermediate Products, Final Products, and Waste Products
Generated During Manufacture and Use:

- Quinone

### Hydrogenation

- See Edible Fat Processing

### Hydronaphthalenes

Raw Materials, Intermediate Products, Final Products, and Waste Products
Generated During Manufacture and Use:

- Naphthalene

### Ice Making

Raw Materials, Intermediate Products, Final Products, and Waste Products
Generated During Manufacture and Use:

- Sulfur dioxide

## Ignition Compounds

Raw Materials, Intermediate Products, Final Products, and Waste Product Generated During Manufacture and Use:

- Phosphorus

## Incandescent Lamps

Raw Materials, Intermediate Products, Final Products, and Waste Product Generated During Manufacture and Use:

- Thorium
- Zirconium

## Incandescent Mantles

Raw Materials, Intermediate Products, Final Products, and Waste Product Generated During Manufacture and Use:

- Thorium

## Incendiaries

Raw Materials, Intermediate Products, Final Products, and Waste Product Generated During Manufacture and Use:

- Magnesium
- Phosphorus

## Incinerators

Raw Materials, Intermediate Products, Final Products, and Waste Products Generated During Manufacture and Use:

- Dioxins
- Garbage
- Heavy metals
- Polychlorinated biphenyls
- Semivolatile compounds
- Tetrahydrofurans

## Indigo

Raw Materials, Intermediate Products, Final Products, and Waste Products Generated During Manufacture and Use:

- Ethylene chlorohydrin

## Indigo, Synthetic

Raw Materials, Intermediate Products, Final Products, and Waste Products Generated During Manufacture and Use:

- Methyl alcohol
- Phthalic anhydride

## Infrared Optical Instruments

General Types of Associated Materials:

- Glass

Raw Materials, Intermediate Products, Final Products, and Waste Product Generated During Manufacture and Use:

- Thallium

## Ink

General Types of Associated Materials:

- Antioxidants
- Detergents
- Resins
- Soaps
- Solvents
- Turpentine
- Varnishes

Raw Materials, Intermediate Products, Final Products, and Waste Product Generated During Manufacture and Use:

- Aniline
- Arsenic
- Cerium
- Chromium
- Cobalt
- Dichlorobenzidine
- Ethylene glycol
- Formaldehyde
- Manganese
- Mercury
- Methyl alcohol
- Molybdenum
- Nickel
- Oxalic acid
- Platinum
- Potassium hydroxide
- Silver
- Tin

Other Associated Materials:

- Benzene
- Carbon tetrachloride
- Ethyl alcohol
- Ethylene glycol ether
- Ketones
- Xylenes

## Insecticides

General Types of Associated Materials:

- Kerosene
- Naphtha
- Turpentine

Raw Materials, Intermediate Products, Final Products, and Waste Products Generated During Manufacture and Use:

- Acetic acid
- Aniline dyes
- Arsenic
- Cadmium
- Calcium arsenate
- Calcium oxide
- Carbon tetrachloride
- Chlorinated benzenes
- Copper
- Dichloroethane
- Dioxins
- Ethanolamines
- Ethylenediamine
- Formic acid
- Furans
- Furfural
- Hydrazine and derivatives
- Hydrogen fluoride
- Lead

Raw Materials, Intermediate Products, Final Products, and Waste Products
Generated During Manufacture and Use (cont.):

- Lead arsenate
- Lead arsenite
- Mercury
- Methoxychlor
- n-Butylamine
- Nicotine
- Phenols
- Phosphides
- Phosphorus
- Phthalic anhydride
- Piperonly compounds
- Pyrethrum
- Rotenone
- Selenium
- Strobane
- Sulfur chloride
- Tetrachloroethane
- Thallium
- Tin

## Instrument Cleaning

General Types of Associated Materials:

- Solvents

Other Associated Materials:

- Perchloroethene
- Trichloroethene

## Insulation

Raw Materials, Intermediate Products, Final Products, and Waste Products Generated During Manufacture and Use:

- Chlorinated naphthalenes
- Cresol
- Isocyanates

Other Associated Materials:

- Ethyl benzene

## Iodine

Other Associated Materials:

- Carbon disulfide

## Iron Enameling

Raw Materials, Intermediate Products, Final Products, and Waste Products Generated During Manufacture and Use:

- Hydrogen fluoride

## Iron Galvanizing

Raw Materials, Intermediate Products, Final Products, and Waste Products Generated During Manufacture and Use:

- Hydrogen fluoride
- Zinc

## Iron Refining and Casting

Raw Materials, Intermediate Products, Final Products, and Waste Products Generated During Manufacture and Use:

- Calcium cyanamide
- Chlorine
- Ferric sulfate
- Ferrous sulfate
- Fluorides
- Heavy metals
- Hydrogen fluoride
- Sulfuric acid
- Tellurium

## Isopropyl Derivatives

Raw Materials, Intermediate Products, Final Products, and Waste Products Generated During Manufacture and Use:

- Isopropyl alcohol

## Ivory

Raw Materials, Intermediate Products, Final Products, and Waste Products Generated During Manufacture and Use:

- Silver

## Jet Engines

Raw Materials, Intermediate Products, Final Products, and Waste Products Generated During Manufacture and Use:

- Magnesium
- MOCA
- Titanium

## Jet Fuels

General Types of Associated Materials:

- Kerosene

Raw Materials, Intermediate Products, Final Products, and Waste Products Generated During Manufacture and Use:

- Benzene
- Ethyl benzene
- Hydrazine and derivatives
- Mercaptans
- Polynuclear aromatic hydrocarbons
- Toluene
- Xylenes

## Jewelry

General Types of Associated Materials:

- Acids
- Fluxes
- Polishes
- Resins
- Rouge
- Solders
- Solvents

Raw Materials, Intermediate Products, Final Products, and Waste Products Generated During Manufacture and Use:

- Arsine
- Cadmium
- Chromates
- Chromium
- Cyanides
- Hydrogen cyanide
- Mercury
- Nickel
- Nitrogen
- Platinum
- Silver
- Sulfuric acid
- Thallium

Other Associated Materials:

- Mercury

## Kelthane

General Types of Associated Materials:

- Organochlorine insecticides

Raw Materials, Intermediate Products, Final Products, and Waste Products Generated During Manufacture and Use:

- See Insecticides

## Kepone

General Types of Associated Materials:

- Organochlorine insecticides

Raw Materials, Intermediate Products, Final Products, and Waste Products Generated During Manufacture and Use:

- See Insecticides

## Kerosene

Raw Materials, Intermediate Products, Final Products, and Waste Products Generated During Manufacture and Use:

- Benzene
- Butadiene
- Cyclohexane
- Cyclohexene
- Cyclopropane
- Ethyl benzene
- Methylcyclohexene
- n-Heptane
- n-Hexane
- Naphthalene
- Polynuclear aromatic hydrocarbons
- Toluene
- Xylenes

## Ketones

Raw Materials, Intermediate Products, Final Products, and Waste Products Generated During Manufacture and Use:

- Acetic acid

## Laboratory Wares

General Types of Associated Materials:

- Ceramics

Raw Materials, Intermediate Products, Final Products, and Waste Products
Generated During Manufacture and Use:

- Platinum

## Lacquer

General Types of Associated Materials:

- Acetates
- Resins
- Solvents
- Turpentine
- Waxes

Raw Materials, Intermediate Products, Final Products, and Waste Products
Generated During Manufacture and Use:

- Acetaldehyde
- Antimony
- Cellulose acetate
- Cellulose Esters
- Chlorinated benzenes
- Chlorodiphenyls
- Epichlorohydrin
- Ethyl silicate
- Formic acid
- Isocyanates
- Molybdenum
- n-Hexane
- Titanium
- Zinc

Other Associated Materials:

- Amyl alcohol
- Carbon disulfide
- Carbon tetrachloride
- Chloroform
- Cyclohexane
- Cyclohexene
- Cyclopropane
- Dichloroethyl ether
- Dioxane
- Ethyl alcohol
- Ethyl benzene
- Ethylene chlorohydrin
- Ethylene glycol ether
- Isopropyl alcohol
- Ketones
- Methylcyclohexene
- n-Butyl alcohol
- n-Propyl alcohol
- Tetrachloroethane
- Toluene
- Xylenes

## Lacquer Remover

- See Lacquer

## Lacquer Thinner

- See Lacquer

## Lamp Filaments

Raw Materials, Intermediate Products, Final Products, and Waste Product
Generated During Manufacture and Use:

- Carbon
- Titanium

## Lampblack

Raw Materials, Intermediate Products, Final Products, and Waste Product
Generated During Manufacture and Use:

- Carbon
- Creosote
- Phenols
- Polynuclear aromatic hydrocarbons

## Larvicides

Raw Materials, Intermediate Products, Final Products, and Waste Product:
Generated During Manufacture and Use:

- Arsenates
- Formates

## Lasting

General Types of Associated Materials:

- Glues
- Leather

Other Associated Materials:

- Methyl alcohol

## Latexes

Raw Materials, Intermediate Products, Final Products, and Waste Products
Generated During Manufacture and Use:

- Fluorides
- Formaldehyde

## Laundries

General Types of Associated Materials:

- Alkalis
- Bactericides
- Bleaches
- Brighteners
- Detergents
- Enzymes
- Fungicides
- Soaps
- Surfactants

Raw Materials, Intermediate Products, Final Products, and Waste Products
Generated During Manufacture and Use:

- Acetic acid
- Chlorinated lime
- Chlorine
- Formic acid
- Oxalic acid
- Sodium hydroxide

Other Associated Materials:

- Chlorine
- Ethylene chlorohydrin

## Lead Refining and Processing

Raw Materials, Intermediate Products, Final Products, and Waste Products
Generated During Manufacture and Use:

- Arsenic
- Arsine
- Dibromomethane
- Lead
- Tellurium

## Lead, White

Raw Materials, Intermediate Products, Final Products, and Waste Products
Generated During Manufacture and Use:

- Acetic acid
- Lead

## Leather Processing

Raw Materials, Intermediate Products, Final Products, and Waste Products
Generated During Manufacture and Use:

- Ammonia
- Aniline
- Antimony
- Arsenic
- Benzyl chloride
- Boron
- Cadmium
- Calcium oxide
- Chromium
- Cyanides
- Formaldehyde
- Formic acid
- Lead
- Mercury

Raw Materials, Intermediate Products, Final Products, and Waste Products Generated During Manufacture and Use (cont.):

- Molybdenum
- n-Butylamine
- Naphthalene
- Phenols
- Picric acid
- Sulfur dioxide
- Sulfuric acid
- Xylenes
- Zinc

Other Associated Materials:

- Benzene
- Ethyl benzene
- Hydrogen chloride
- Methyl alcohol
- Methylene chloride
- Perchloroethene
- Toluene
- Xylenes

## Leather, Synthetic

General Types of Associated Materials:

- Acetates

Raw Materials, Intermediate Products, Final Products, and Waste Products Generated During Manufacture and Use:

- Ketones
- Methyl alcohol
- n-Butyl alcohol
- Tetrachloroethane

## Lemonade

Raw Materials, Intermediate Products, Final Products, and Waste Product Generated During Manufacture and Use:

- Formates

## Lighter Flints

Raw Materials, Intermediate Products, Final Products, and Waste Product Generated During Manufacture and Use:

- Cerium

## Lignin Extraction

Other Associated Materials:

- Ethylene chlorohydrin

## Liniments

Other Associated Materials:

- Isopropyl alcohol

## Linoleum Manufacturing

General Types of Associated Materials:

- Dyes
- Oils
- Pigments
- Resins
- Solvents

Raw Materials, Intermediate Products, Final Products, and Waste Products
Generated During Manufacture and Use:

- Asbestos
- Asphalt
- Polynuclear aromatic hydrocarbons

Other Associated Materials:

- Benzene

## Linotype Metal

Raw Materials, Intermediate Products, Final Products, and Waste Products
Generated During Manufacture and Use:

- Antimony
- Arsenic
- Tin

## Liquid Soaps

Raw Materials, Intermediate Products, Final Products, and Waste Products
Generated During Manufacture and Use:

- Ethylene glycol ether

## Liquor, Distilled

General Types of Associated Materials:

- Alcohols

Raw Materials, Intermediate Products, Final Products, and Waste Product
Generated During Manufacture and Use:

- Pyridine

## Lithography

General Types of Associated Materials:

- Ink
- Solvents
- Turpentine
- Varnishes

Raw Materials, Intermediate Products, Final Products, and Waste Product
Generated During Manufacture and Use:

- Aniline
- Chromium
- Copper
- Hydrogen sulfide
- Nitrogen
- Phosphoric acid
- Potassium hydroxide

## Lithopone

Raw Materials, Intermediate Products, Final Products, and Waste Product
Generated During Manufacture and Use:

- Hydrogen sulfide

## Lounges

• See Bars, Clubs, and Lounges

## Lubricant Additives

Raw Materials, Intermediate Products, Final Products, and Waste Products Generated During Manufacture and Use:

• Arsenic
• Calcium oxide
• Chlorodiphenyls
• Ethanolamines
• Graphite
• Ketones
• Lead
• Molybdenum
• Nickel
• Nitrosodimethylamine
• Phosphorus
• Tetramethylthiuram disulfide
• Tricresyl phosphates

## Lucifers

Raw Materials, Intermediate Products, Final Products, and Waste Products Generated During Manufacture and Use:

• Phosphorus

## Lysine

Raw Materials, Intermediate Products, Final Products, and Waste Products Generated During Manufacture and Use:

• Furfural

## Maneb

General Types of Associated Materials:

- Dithiocarbamate fungicides

Raw Materials, Intermediate Products, Final Products, and Waste Product Generated During Manufacture and Use:

- See Fungicides


## Machining

- See Metal Working and Machining


## Magnetic Alloys

Raw Materials, Intermediate Products, Final Products, and Waste Product Generated During Manufacture and Use:

- Chromium
- Iron
- Nickel


## Magnetic Tapes

General Types of Associated Materials:

- Cellulose
- Polyester

Raw Materials, Intermediate Products, Final Products, and Waste Product Generated During Manufacture and Use:

- Nickel

## Magnets

Raw Materials, Intermediate Products, Final Products, and Waste Products
Generated During Manufacture and Use:

- Bismuth
- Iron
- Nickel

## Malathion

General Types of Associated Materials:

- Organophosphate insecticides

Raw Materials, Intermediate Products, Final Products, and Waste Products
Generated During Manufacture and Use:

- See Insecticides

## Maleic Anhydride

Raw Materials, Intermediate Products, Final Products, and Waste Products
Generated During Manufacture and Use:

- Benzene

## Malonic Acid

Raw Materials, Intermediate Products, Final Products, and Waste Products
Generated During Manufacture and Use:

- Ethylene chlorohydrin

## Malt Beverages

General Types of Associated Materials:

- Alcohols

Raw Materials, Intermediate Products, Final Products, and Waste Product Generated During Manufacture and Use:

- Arsenic

## Masonry

- See Bricks and Masonry

## Masonry Cleaning

Raw Materials, Intermediate Products, Final Products, and Waste Product: Generated During Manufacture and Use:

- Hydrogen fluoride
- Muriatic acid

## Masonry Preservatives

General Types of Associated Materials:

- Silicone
- Waterproofing

Raw Materials, Intermediate Products, Final Products, and Waste Product: Generated During Manufacture and Use:

- Ethyl silicate

## Match Manufacturing

General Types of Associated Materials:

- Dyes
- Glues
- Gums
- Waxes

Raw Materials, Intermediate Products, Final Products, and Waste Products Generated During Manufacture and Use:

- Antimony
- Chromates
- Dextrins
- Formaldehyde
- Graphite
- Lead
- Manganese
- Phosphorus
- Picric acid
- Potassium chlorate

Other Associated Materials:

- Ammonium compounds

## MCPA

General Types of Associated Materials:

- Chlorinated herbicides

Raw Materials, Intermediate Products, Final Products, and Waste Products Generated During Manufacture and Use:

- See Herbicides

## MCPP

General Types of Associated Materials:

- Chlorinated herbicides

Raw Materials, Intermediate Products, Final Products, and Waste Product Generated During Manufacture and Use:

- See Herbicides

## Meat Packing

- See Butchers

## Mechanics

- See Aircraft Manufacturing
- See Automotive Service Stations

## Medical Facilities

General Types of Associated Materials:

- Adhesives and removers
- Antiseptics
- Detergents
- Diesel fuel
- Disinfectants
- Drugs
- Gasoline
- Heating oil
- Rubber
- Soaps
- Stains
- Waxes

Raw Materials, Intermediate Products, Final Products, and Waste Products
Generated During Manufacture and Use:

- Aniline dyes
- Bacteria
- Benzol
- Chromium
- Ethyl chloride
- Formaldehyde
- Mercury
- Osmium tetroxide
- Picric acid
- Propyl alcohol
- Tetrachloroethene
- Toluene
- Viruses

Other Associated Materials:

- Dioxane
- Ethyl alcohol
- Xylenes

## Medical Laboratories

Raw Materials, Intermediate Products, Final Products, and Waste Products
Generated During Manufacture and Use:

- Benzidine
- Formaldehyde

## Melamine Resins

Raw Materials, Intermediate Products, Final Products, and Waste Products
Generated During Manufacture and Use:

- Formaldehyde

## Mercerizers

• See Cloth Processing

## Mercury Vapor Lamps

Raw Materials, Intermediate Products, Final Products, and Waste Product;
Generated During Manufacture and Use:

• Mercury

## Merphos

General Types of Associated Materials:

• Organophosphorus insecticides

Raw Materials, Intermediate Products, Final Products, and Waste Product;
Generated During Manufacture and Use:

• See Insecticides

## Metal Cleaning

• See Plating and Polishing

## Metal Forging

General Types of Associated Materials:

• Lubricants
• Paints
• Solvents
• Waste oils

Raw Materials, Intermediate Products, Final Products, and Waste Products
Generated During Manufacture and Use:

- Arsenic
- Cadmium
- Chromium
- Copper
- Cyanides
- Ethyl silicate
- Heavy metals
- Lead
- Nickel
- Phenols
- Phosphorus
- Polychlorinated biphenyls
- Polynuclear aromatic hydrocarbons
- Zinc

Other Associated Materials:

- Benzene
- Chloroform
- Cyclohexane
- Ethyl benzene
- Hexane
- Methylene chloride
- Naphthalene
- Perchloroethene
- Toluene
- Trichloroethane
- Trichloroethene
- Vinyl chloride
- Xylenes

## Metal Parts Manufacturing

General Types of Associated Materials:

- Cutting oils
- Grease
- Petroleum fuels
- Waste oils

Raw Materials, Intermediate Products, Final Products, and Waste Products Generated During Manufacture and Use:

- Aluminum
- Cadmium
- Chromium
- Cyanides
- Lead
- Nickel
- Polychlorinated biphenyls

Other Associated Materials:

- Benzene
- Dichlorobenzene
- Ethyl benzene
- Perchloroethene
- Toluene
- Trichloroethane
- Trichloroethene
- Xylenes

## Metal Polishing

General Types of Associated Materials:

- Abrasives
- Acids
- Alkalis
- Chlorinated waxes
- Degreasers
- Detergents
- Gasoline
- Metal cleaners
- Naphtha
- Oils
- Soaps
- Soluble oils
- Solvents
- Waste oils

Raw Materials, Intermediate Products, Final Products, and Waste Products
Generated During Manufacture and Use:

- Aluminum
- Arsenic
- Asbestos
- Cadmium
- Chlorides
- Chromic acid
- Chromium
- Copper
- Cyanides
- Ethylene glycol
- Heavy metals
- Iron
- Lead
- Nickel
- Nitrates
- Polychlorinated biphenyls
- Sulfate
- Sulfuric acid
- Triethanolamine
- Zinc

Other Associated Materials:

- Ammonia
- Benzene
- Chloroform
- Chromic acid
- Ethyl benzene
- Hexane
- Perchloroethene
- Toluene
- Trichloroethane
- Trichloroethene
- Xylenes

## Metal Stamping

General Types of Associated Materials:

- Oils
- Paints
- Petroleum fuels
- Solvents
- Waste oils

Raw Materials, Intermediate Products, Final Products, and Waste Product Generated During Manufacture and Use:

- Arsenic
- Cadmium
- Chromium
- Copper
- Cyanides
- Heavy metals
- Lead
- Nickel
- Phenols
- Phosphorus
- Polychlorinated biphenyls
- Zinc

Other Associated Materials:

- Benzene
- Chloroform
- Cyclohexane
- Ethyl benzene
- Hexane
- Methylene chloride
- Naphthalene
- Perchloroethene
- Polynuclear aromatic hydrocarbons
- Toluene
- Trichloroethane
- Trichloroethene
- Vinyl chloride
- Xylenes

## Metal Working and Machining

General Types of Associated Materials:

- Antioxidants
- Cutting fluids, aqueous
- Cutting oils
- Cutting oils, insoluble
- Germicides
- Lubricants
- Rust inhibitors
- Soluble oils
- Solvents

Raw Materials, Intermediate Products, Final Products, and Waste Products Generated During Manufacture and Use:

- Chromates
- Zinc

Other Associated Materials:

- Chloroform
- Dichloroethane
- Dichloroethene
- Methylene chloride
- Trichloroethene
- Vinyl chloride

## Methacrylates

Raw Materials, Intermediate Products, Final Products, and Waste Products Generated During Manufacture and Use:

- Methyl alcohol

## Methionine

Raw Materials, Intermediate Products, Final Products, and Waste Product
Generated During Manufacture and Use:

- Acrolein
- Mercaptans

## Methoxychlor

General Types of Associated Materials:

- Organochlorine insecticides

Raw Materials, Intermediate Products, Final Products, and Waste Product:
Generated During Manufacture and Use:

- See Insecticides

## Methyl Alcohol

Raw Materials, Intermediate Products, Final Products, and Waste Product:
Generated During Manufacture and Use:

- Oxalic acid

## Methyl Amines

Raw Materials, Intermediate Products, Final Products, and Waste Product:
Generated During Manufacture and Use:

- Methyl alcohol

### Methyl Ethyl Ketone

• See Ketones

### Methyl Halides

Raw Materials, Intermediate Products, Final Products, and Waste Products
Generated During Manufacture and Use:

• Methyl alcohol

### Methyl Isobutyl Ketone

• See Ketones

### Methyl n-Butyl Ketone

• See Ketones

### Methyl n-Propyl Ketone

• See Ketones

### Methylating Agents

Raw Materials, Intermediate Products, Final Products, and Waste Products
Generated During Manufacture and Use:

• Dimethyl sulfate

## Methylenebis(2-Chloroaniline)

• See MOCA

## Mevinphos

General Types of Associated Materials:

• Organophosphorus insecticides

Raw Materials, Intermediate Products, Final Products, and Waste Products
Generated During Manufacture and Use:

• See Insecticides

## Mildew-proofing

Raw Materials, Intermediate Products, Final Products, and Waste Products
Generated During Manufacture and Use:

• Mercury

## Military Hardware

General Types of Associated Materials:

• Explosives
• Lubricants
• Metals
• Solvents

Raw Materials, Intermediate Products, Final Products, and Waste Products
Generated During Manufacture and Use:

• Beryllium
• Heavy metals

Other Associated Materials:

- Benzene
- Chlorobenzene
- Chloroform
- Ethyl benzene
- Tetrachloroethene
- Toluene
- Trichloroethane
- Trichloroethene
- Xylenes

## Milk Processing and Distribution

General Types of Associated Materials:

- Bacteria
- Bactericides
- Deodorants
- Detergents
- Viruses

Other Associated Materials:

Polybrominated biphenyls

## Mineralogical Laboratories

General Types of Associated Materials:

- Abrasives
- Glues
- Oils
- Polishes
- Resins
- Solvents

## Mining

General Types of Associated Materials:

- Acids
- Explosives
- Lubricants
- Metals

Raw Materials, Intermediate Products, Final Products, and Waste Products Generated During Manufacture and Use:

- Amyl alcohol
- Antimony
- Benzene
- Chlorides
- Cyanides
- Ethyl benzene
- Heavy metals
- Nitrates
- Polynuclear aromatic hydrocarbons
- Radiation, alpha
- Radiation, beta
- Radiation, gamma
- Sodium
- Sulfates
- Toluene
- Trinitrotoluene
- Xylenes

## Mirex

General Types of Associated Materials:

- Organochlorine insecticides

Raw Materials, Intermediate Products, Final Products, and Waste Products Generated During Manufacture and Use:

- See Insecticides

## Mirror Making

General Types of Associated Materials:

- Acids
- Lacquers
- Solvents
- Varnishes

Raw Materials, Intermediate Products, Final Products, and Waste Products Generated During Manufacture and Use:

- Acetaldehyde
- Ammonia
- Cyanides
- Formaldehyde
- Platinum
- Silver
- Tartaric acid

Other Associated Materials:

- Ammonia

## Molluscicides

Raw Materials, Intermediate Products, Final Products, and Waste Products Generated During Manufacture and Use:

- Copper

## Monel-metal

Raw Materials, Intermediate Products, Final Products, and Waste Products Generated During Manufacture and Use:

- Nickel

## Mordants

General Types of Associated Materials:

- Dyes
- Etching

Raw Materials, Intermediate Products, Final Products, and Waste Product Generated During Manufacture and Use:

- Amyl alcohol
- Antimony
- Copper
- Molybdenum
- Potassium hydroxide
- Tin
- Titanium
- Vanadium

## Morticians

- See Embalming

## Moth Preventatives

Raw Materials, Intermediate Products, Final Products, and Waste Products Generated During Manufacture and Use:

- Camphor
- Chlorinated benzenes
- Naphthalene

## Moth Repellant

Raw Materials, Intermediate Products, Final Products, and Waste Products Generated During Manufacture and Use:

- Naphthalene

## Mouth Wash

Other Associated Materials:

- Isopropyl alcohol

## Muriatic Acid

Other Associated Materials:

- Hydrogen chloride

## Mustard Gas

General Types of Associated Materials:

- Solvents

Raw Materials, Intermediate Products, Final Products, and Waste Products Generated During Manufacture and Use:

- Chlorinated lime

## Mylar

Raw Materials, Intermediate Products, Final Products, and Waste Products Generated During Manufacture and Use:

- Phthalic anhydride

## Nabam

General Types of Associated Materials:

- Dithiocarbamate fungicides

Raw Materials, Intermediate Products, Final Products, and Waste Products Generated During Manufacture and Use:

- See Fungicides

## Nail Polishes

Raw Materials, Intermediate Products, Final Products, and Waste Products Generated During Manufacture and Use:

- Epichlorohydrin

Other Associated Materials:

- Ethylene glycol ether

## Naled

General Types of Associated Materials:

- Organophosphate insecticides

Raw Materials, Intermediate Products, Final Products, and Waste Products
Generated During Manufacture and Use:

- See Insecticides

## Naphtha

Raw Materials, Intermediate Products, Final Products, and Waste Products
Generated During Manufacture and Use:

- Benzene
- Cumenes
- Ethyl benzene
- Polynuclear aromatic hydrocarbons
- Toluene
- Xylenes

## Naphthionic Acid

Raw Materials, Intermediate Products, Final Products, and Waste Products
Generated During Manufacture and Use:

- alpha-Naphthylamine

## Naphthols

Raw Materials, Intermediate Products, Final Products, and Waste Products
Generated During Manufacture and Use:

- Naphthalene

## Naphthylamines

Raw Materials, Intermediate Products, Final Products, and Waste Products Generated During Manufacture and Use:

- Naphthalene

## Naphthalene Processing

Other Associated Materials:

- Dichloroethyl ether

## Natural Gas Drilling

- See Petroleum Exploration

## Natural Gas Production

Raw Materials, Intermediate Products, Final Products, and Waste Products Generated During Manufacture and Use:

- Benzene
- Brine
- Ethanolamines
- Hydrogen sulfide

Raw Materials, Intermediate Products, Final Products, and Waste Products
Generated During Manufacture and Use (cont.):

- Toluene
- Zinc


## Natural Gas Refining

Raw Materials, Intermediate Products, Final Products, and Waste Products
Generated During Manufacture and Use:

- Benzene
- Toluene


## Natural Gas Storage

- See Petroleum Exploration


## Nematocides

Raw Materials, Intermediate Products, Final Products, and Waste Products
Generated During Manufacture and Use:

- Cadmium
- Nitrosodimethylamine


## Neoprene

Raw Materials, Intermediate Products, Final Products, and Waste Products
Generated During Manufacture and Use:

- Chloroprene

## Neutron Absorbers

Raw Materials, Intermediate Products, Final Products, and Waste Products Generated During Manufacture and Use:

- Cadmium

## Neville's Acid

Raw Materials, Intermediate Products, Final Products, and Waste Products Generated During Manufacture and Use:

- alpha-Naphthylamine

## Nickel

Raw Materials, Intermediate Products, Final Products, and Waste Products Generated During Manufacture and Use:

- Cobalt

Other Associated Materials:

- Fluorides

## Nickeling

- See Plating and Polishing

### Nitraniline

Raw Materials, Intermediate Products, Final Products, and Waste Products Generated During Manufacture and Use:

- Aniline

### Nitrates

Raw Materials, Intermediate Products, Final Products, and Waste Products Generated During Manufacture and Use:

- Nitrogen

### Nitric Acid

Raw Materials, Intermediate Products, Final Products, and Waste Products Generated During Manufacture and Use:

- Nitrogen

### Nitriles

Raw Materials, Intermediate Products, Final Products, and Waste Products Generated During Manufacture and Use:

- Acrylonitrile
- Hydrogen cyanide

## Nitrites

Raw Materials, Intermediate Products, Final Products, and Waste Products Generated During Manufacture and Use:

- Nitrogen

## Nitrobenzene

Raw Materials, Intermediate Products, Final Products, and Waste Products Generated During Manufacture and Use:

- Benzene

## Nitrocellulose

General Types of Associated Materials:

- Acetates
- Cellulose

Raw Materials, Intermediate Products, Final Products, and Waste Products Generated During Manufacture and Use:

- Isopropyl alcohol

Other Associated Materials:

- Ethyl alcohol
- Ethyl ether
- Ketones
- Tricresyl phosphates

## Nitrogen Processing

Raw Materials, Intermediate Products, Final Products, and Waste Products Generated During Manufacture and Use:

- Calcium cyanamide

## Nitroglycerine

Raw Materials, Intermediate Products, Final Products, and Waste Products Generated During Manufacture and Use:

- Ethylene glycol dinitrate
- Nitric acid
- Nitroglycerin
- Sodium carbonate
- Sulfuric acid

## Nitrotoluene

Raw Materials, Intermediate Products, Final Products, and Waste Products Generated During Manufacture and Use:

- Toluene

## Nitrous Acid

Raw Materials, Intermediate Products, Final Products, and Waste Products Generated During Manufacture and Use:

- Nitrogen

## Novocaine

Raw Materials, Intermediate Products, Final Products, and Waste Product Generated During Manufacture and Use:

- Ethylene chlorohydrin

## Nuclear Fuel

Raw Materials, Intermediate Products, Final Products, and Waste Product Generated During Manufacture and Use:

- Radiation, alpha
- Radiation, beta
- Radiation, gamma
- Plutonium
- Thorium
- Uranium

## Nuclear Instruments

Raw Materials, Intermediate Products, Final Products, and Waste Products Generated During Manufacture and Use:

- Boron

## Nuclear Reactors

Raw Materials, Intermediate Products, Final Products, and Waste Products Generated During Manufacture and Use:

- Beryllium
- Graphite
- Thorium
- Titanium
- Uranium

## Nuts

Raw Materials, Intermediate Products, Final Products, and Waste Products Generated During Manufacture and Use:

- Tetramethylthiuram disulfide

## Nutylaminophenol

Raw Materials, Intermediate Products, Final Products, and Waste Products Generated During Manufacture and Use:

- n-Butylamine

## Nylon

Raw Materials, Intermediate Products, Final Products, and Waste Products Generated During Manufacture and Use:

- Trichloroethane
- Furfural

Other Associated Materials:

- Cyclohexane
- Cyclohexene
- Cyclopropane
- Methylcyclohexene

## Odorants

General Types of Associated Materials:

- Acetates
- Mercaptans

## Office Supply Manufacturing

General Types of Associated Materials:

- Adhesives and removers
- Duplicating fluids
- Duplicating materials
- Ink removers
- Inks
- Rubber
- Solvents
- Typewriter ribbons

## Oil and Gas Extraction

- See Petroleum Exploration

## Oil and Grease Processing

General Types of Associated Materials:

- Metals
- Petroleum hydrocarbons
- Solvents

## Oil Drilling

- See Petroleum Exploration

## Oils, Mineral

Other Associated Materials:

- Dimethyl sulfate

## Olefins

Raw Materials, Intermediate Products, Final Products, and Waste Products Generated During Manufacture and Use:

- Boron hydrides

Other Associated Materials:

- Hydrogen chloride

## Optical Lenses

General Types of Associated Materials:

- Glass
- Polishes

Raw Materials, Intermediate Products, Final Products, and Waste Products Generated During Manufacture and Use:

- Thallium

## Optical Whitening Agents

Raw Materials, Intermediate Products, Final Products, and Waste Products Generated During Manufacture and Use:

- Aniline

## Orange Packing

Raw Materials, Intermediate Products, Final Products, and Waste Products Generated During Manufacture and Use:

- Diphenyl

## Ore Flotation Agents

Raw Materials, Intermediate Products, Final Products, and Waste Products Generated During Manufacture and Use:

- Amyl alcohol
- Cresol

## Ore Smelting and Processing

General Types of Associated Materials:

- Metals

Raw Materials, Intermediate Products, Final Products, and Waste Products Generated During Manufacture and Use:

- Cresol
- Dichloroethane
- Sulfur dioxide

Other Associated Materials:

- Hydrogen chloride

## Organic Chemical Synthesis

General Types of Associated Materials:

- Acetates

Raw Materials, Intermediate Products, Final Products, and Waste Products Generated During Manufacture and Use:

- Acridine
- Acrylonitrile
- Allyl alcohol

Raw Materials, Intermediate Products, Final Products, and Waste Products
Generated During Manufacture and Use (cont.):

- alpha-Naphthylamine
- Ammonia
- Benzidine
- bis(chloromethyl) Ether
- Boron hydrides
- Bromine
- Chlorinated benzenes
- Chloromethyl methyl ether
- Chromium
- Creosote
- Dinitrobenzene
- Dinitrophenol
- Dinitrotoluene
- Diphenyl
- Epichlorohydrin
- Ethyl alcohol
- Ethyl benzene
- Ethylene chlorohydrin
- Ethylene oxide
- Formates
- Furfural
- Hydrogen cyanide
- Hydroquinone
- Isocyanates
- Magnesium
- Methyl alcohol
- Nitrobenzene
- Nitrogen
- Nitroparaffins
- Nitrophenol
- Phenols
- Phthalic anhydride
- Propylene dichloride
- Pyridine
- Quinone
- Selenium
- Sulfur chloride
- Tetrachloroethene

Raw Materials, Intermediate Products, Final Products, and Waste Products Generated During Manufacture and Use (cont.):

- Thallium
- Thorium
- Vinyl chloride

Other Associated Materials:

- Dimethylformamide
- Fluorides
- Trichloroethane

## Oxalates

Raw Materials, Intermediate Products, Final Products, and Waste Products Generated During Manufacture and Use:

- Formic acid
- Oxalic acid

## Oxalic Acid

Raw Materials, Intermediate Products, Final Products, and Waste Products Generated During Manufacture and Use:

- Nitrogen

## Oxidizing Agents

Raw Materials, Intermediate Products, Final Products, and Waste Products Generated During Manufacture and Use:

- Bromine

## Paint

General Types of Associated Materials:

- Acetates
- Acids
- Acrylics
- Alkalis
- Antimildew agents
- Coal tar products
- Cutting oils
- Driers
- Enamels
- Epoxy resins
- Fish oils
- Latex
- Naphtha
- Paint removers
- Paints
- Pigments
- Plasticizers
- Resins
- Solvents
- Thinners
- Turpentine

Raw Materials, Intermediate Products, Final Products, and Waste Products Generated During Manufacture and Use:

- Aluminum
- Antimony
- Arsenic
- Boron
- Cadmium
- Chlorinated benzenes
- Chromates
- Chromium
- Cobalt
- Copper
- Epichlorohydrin
- Ethyl silicate

Raw Materials, Intermediate Products, Final Products, and Waste Products
Generated During Manufacture and Use (cont.):

- Ethylene glycol
- Graphite
- Hydroquinone
- Isocyanates
- Lead
- Manganese
- Mercury
- Molybdenum
- Nickel
- Oxalic acid
- Phenols
- Phthalates
- Potassium hydroxide
- Pyridine
- Sulfuric acid
- Tin
- Titanium
- Toluene
- Zinc

Other Associated Materials:

- Acetone
- Amyl alcohol
- Benzene
- Carbon disulfide
- Dichloroethane
- Dichloroethyl ether
- Dioxane
- Ethyl benzene
- Ethyl ether
- Ethylene glycol ether
- Hydrogen chloride
- Ketones
- Methyl alcohol
- Nitrobenzene
- Perchloroethene
- Polynuclear aromatic hydrocarbons

Other Associated Materials (cont.):

- Pyrene
- Tetrachloroethane
- Toluene
- Trichloroethane
- Trichloroethene
- Xylenes

## Paint Removers

Raw Materials, Intermediate Products, Final Products, and Waste Products Generated During Manufacture and Use:

- Cresol
- Oxalic acid
- Phenols
- Potassium hydroxide

Other Associated Materials:

- Amyl alcohol
- Cyclohexane
- Cyclohexene
- Cyclopropane
- Dichloroethyl ether
- Dioxane
- Ketones
- Methyl alcohol
- Methylcyclohexene
- Methylene chloride

## Paint Thinners

General Types of Associated Materials:

- Mineral spirits
- Naphtha

Other Associated Materials:

- Aliphatic hydrocarbons
- Benzene
- Ethyl benzene
- Nitrobenzene
- Toluene
- Xylenes

## Paper

General Types of Associated Materials:

- Antiflame agents
- Bleaches
- Dyes
- Glues
- Grease
- Lubricants
- Resins
- Waxes

Raw Materials, Intermediate Products, Final Products, and Waste Product Generated During Manufacture and Use:

- Ammonia
- Arsine
- Benzidine
- Chlorinated lime
- Chlorodiphenyls
- Dioxins
- Ethyleneimine
- Formaldehyde
- Formic acid
- Iron
- Ketones
- Lead
- Lignins
- Methyl alcohol
- Mildewproofing

Raw Materials, Intermediate Products, Final Products, and Waste Products
Generated During Manufacture and Use (cont.):

- Oxalic acid
- Phenols
- Phthalates
- Sodium hydroxide
- Sulfuric acid
- Zinc

Other Associated Materials:

- Ethyl benzene
- Titanium
- Trichloroethane
- Xylenes

## Paraldehyde

Raw Materials, Intermediate Products, Final Products, and Waste Products
Generated During Manufacture and Use:

- Acetaldehyde

## Paraquat

General Types of Associated Materials:

- Bipyridyl herbicides

Raw Materials, Intermediate Products, Final Products, and Waste Products
Generated During Manufacture and Use:

- See Herbicides

## Parathion

General Types of Associated Materials:

- Organophosphate insecticides

Raw Materials, Intermediate Products, Final Products, and Waste Produc
Generated During Manufacture and Use:

- See Insecticides

## Parathion Methyl

General Types of Associated Materials:

- Organophosphorus insecticides

Raw Materials, Intermediate Products, Final Products, and Waste Produc
Generated During Manufacture and Use:

- See Insecticides

## Paris Green

Raw Materials, Intermediate Products, Final Products, and Waste Produc
Generated During Manufacture and Use:

- Acetic acid

## Paraffin

General Types of Associated Materials:

- Petroleum hydrocarbons
- Solvents

Raw Materials, Intermediate Products, Final Products, and Waste Products Generated During Manufacture and Use:

- Polynuclear aromatic hydrocarbons


## Pearls, Synthetic

Raw Materials, Intermediate Products, Final Products, and Waste Products Generated During Manufacture and Use:

- Dichloroethene
- Lead
- Tetrachloroethane


## Pencil Manufacturing

General Types of Associated Materials:

- Glues
- Gums
- Lacquer thinners
- Lacquers
- Resins
- Solvents
- Waxes

Raw Materials, Intermediate Products, Final Products, and Waste Products Generated During Manufacture and Use:

- Aniline dyes
- Chromium
- Graphite
- Methyl violet
- Pyridine

## Penicillin

Raw Materials, Intermediate Products, Final Products, and Waste Product Generated During Manufacture and Use:

- Chloroform

## Pentaerythritol

Raw Materials, Intermediate Products, Final Products, and Waste Product Generated During Manufacture and Use:

- Acetaldehyde

## Perchloroethene

- See Tetrachloroethene

## Perfume Manufacturing

General Types of Associated Materials:

- Acetates
- Paraffin

Raw Materials, Intermediate Products, Final Products, and Waste Product Generated During Manufacture and Use:

- Acetaldehyde
- Acrolein
- Aniline
- Benzyl chloride
- Dimethyl sulfate
- Ethyl alcohol
- Ethyl chloride

Raw Materials, Intermediate Products, Final Products, and Waste Products Generated During Manufacture and Use (cont.):

- Ethylene glycol ether
- Formic acid
- Isopropyl alcohol
- Phenols
- Propyl alcohol
- Trichloroethene
- Xylenes

Other Associated Materials:

- Acetic anhydride
- Amyl alcohol
- Cyclohexane
- Ethyl ether
- Toluene

## Pesticides

Raw Materials, Intermediate Products, Final Products, and Waste Products Generated During Manufacture and Use:

- Acetylaminofluorene
- Acrylonitrile
- Ammonia
- Calcium cyanamide
- Chlorinated lime
- Chlorine
- Dinitro-o-cresol
- Dinitrophenol
- Fluorides
- Formates
- Hydrazine and derivatives
- Mercaptans
- n-Butylamine
- Pentachlorophenol
- Phenols

Raw Materials, Intermediate Products, Final Products, and Waste Products
Generated During Manufacture and Use (cont.):

- Phosphorus
- Trichloroethene

Other Associated Materials:

- Ethylene chlorohydrin

## Petroleum Exploration and Production

General Types of Associated Materials:

- Acids
- Alkalis
- Brine
- Crude oil
- Herbicides
- Lubricants
- Paints
- Paraffin
- Pesticides
- Petroleum hydrocarbons
- Solvents

Raw Materials, Intermediate Products, Final Products, and Waste Products
Generated During Manufacture and Use:

- Barium
- Benzene
- Bromide
- Calcium
- Calcium oxide
- Heavy metals
- Hydrogen sulfide
- Iodide
- Polychlorinated biphenyls
- Polynuclear aromatic hydrocarbons

Raw Materials, Intermediate Products, Final Products, and Waste Products Generated During Manufacture and Use (cont.):

- Potassium
- Sodium chloride
- Zinc

Other Associated Materials:

- Benzene
- Ethyl benzene
- Hydrogen chloride
- Toluene
- Xylenes

## Petroleum Refining

General Types of Associated Materials:

- Acids
- Alkalis
- Gasoline
- Kerosene
- Lubricants
- Naphtha
- Paints
- Paraffin
- Pesticides
- Petroleum fuels
- Petroleum hydrocarbons
- Solvents
- Tar and derivatives
- Waxes

Raw Materials, Intermediate Products, Final Products, and Waste Products Generated During Manufacture and Use:

- Aluminum
- Aluminum salts

Raw Materials, Intermediate Products, Final Products, and Waste Products Generated During Manufacture and Use (cont.):

- Ammonia
- Arsenic
- Bacteria
- Benzene
- Bromine
- Chlorides
- Chlorinated naphthalenes
- Chromium
- Copper
- Cyanides
- Dibromoethane
- Ethyl benzene
- Ethylene chlorohydrin
- Furfural
- Heavy metals
- Hydrofluoric acid
- Hydrogen sulfide
- Lead
- Mercaptans
- Mercury
- Molybdenum
- n-Butylamine
- n-Heptane
- n-Hexane
- Phenols
- Polychlorinated biphenyls
- Sodium hydroxide
- Sulfur dioxide
- Toluene
- Zinc

Other Associated Materials:

- Amyl alcohol
- Benzene
- Dichloroethane
- Dimethylformamide
- Ethyl benzene

Other Associated Materials (cont.):

- Hydrogen fluoride
- Methylene chloride
- Toluene
- Xylenes

## Petroleum Bulk Storage

General Types of Associated Materials:

- Diesel fuel
- Gasoline
- Heating oil
- Jet fuel
- Petroleum fuels
- Waste oils

Raw Materials, Intermediate Products, Final Products, and Waste Products Generated During Manufacture and Use:

- Chlorides
- Lead
- Polychlorinated biphenyls

Other Associated Materials:

- Benzene
- Dichloroethane
- Ethyl benzene
- Methyl tertiary butyl ether
- Naphthalene
- Perchloroethene
- Polynuclear aromatic hydrocarbons
- Toluene
- Trimethyl benzene
- Xylenes

## Pewter

Raw Materials, Intermediate Products, Final Products, and Waste Products
Generated During Manufacture and Use:

- Antimony
- Tin

## Pharmaceuticals

General Types of Associated Materials:

- Alcohol
- Acetates
- Solvents

Raw Materials, Intermediate Products, Final Products, and Waste Products
Generated During Manufacture and Use:

- Acetic acid
- Acetonitrile
- Acridine
- Ammonia
- Aniline
- Antimony
- Arsenic
- Benzene
- Benzyl chloride
- Bromine
- Chloroform
- Dichloroethane
- Dichloroethene
- Ethyl alcohol
- Ethylene glycol dinitrate
- Ethylenediamine
- Hexamethylenetetramine
- Ketones
- Manganese
- Mercaptans

Raw Materials, Intermediate Products, Final Products, and Waste Products Generated During Manufacture and Use (cont.):

- Mercury
- n-Butylamine
- Nitrogen
- Nitroglycerin
- Phenols
- Phosphoric acid
- Picric acid
- Sulfur chloride
- Sulfuric acid
- Xylenes
- Zinc

Other Associated Materials:

- Acetic anhydride
- Dimethylformamide
- Hydrogen chloride

## Pharmacies

General Types of Associated Materials:

- Acids
- Alkalis
- Bleaches
- Detergents
- Drugs
- Soaps

Raw Materials, Intermediate Products, Final Products, and Waste Products Generated During Manufacture and Use:

- Iodoform

## Phenol Derivatives

Raw Materials, Intermediate Products, Final Products, and Waste Products Generated During Manufacture and Use:

- Dimethyl sulfate
- Phenols

## Phenolformaldehyde Resins

Raw Materials, Intermediate Products, Final Products, and Waste Products Generated During Manufacture and Use:

- Formaldehyde
- Phenols
- Resins

## Phenolic Plastics

General Types of Associated Materials:

- Phenols

Raw Materials, Intermediate Products, Final Products, and Waste Products Generated During Manufacture and Use:

- Furfural

## Phenolic Resins

General Types of Associated Materials:

- Phenols

Raw Materials, Intermediate Products, Final Products, and Waste Products
Generated During Manufacture and Use:

- Acetaldehyde
- Formic acid

## Phenolics

General Types of Associated Materials:

- Acetates
- Phenols

## Phenols

Raw Materials, Intermediate Products, Final Products, and Waste Products
Generated During Manufacture and Use:

- Benzene
- Formaldehyde
- Sulfuric acid
- Toluene
- Tricresyl phosphates

## Phorate

General Types of Associated Materials:

- Organophosphorus insecticides

Raw Materials, Intermediate Products, Final Products, and Waste Products
Generated During Manufacture and Use:

- See Insecticides

## Phosphate Salts

Raw Materials, Intermediate Products, Final Products, and Waste Product Generated During Manufacture and Use:

* Phosphoric acid

## Phosphoric Acid

Raw Materials, Intermediate Products, Final Products, and Waste Product Generated During Manufacture and Use:

* Nitrogen
* Phosphorus
* Sulfuric acid

## Phosphorus

Raw Materials, Intermediate Products, Final Products, and Waste Product Generated During Manufacture and Use:

* Cerium
* Ethyl chloride
* Germanium
* Tetrachloroethane
* Thallium

Other Associated Materials:

* Carbon disulfide

### Photocells

Raw Materials, Intermediate Products, Final Products, and Waste Products Generated During Manufacture and Use:

- Cadmium
- Selenium
- Silicon

### Photocopying

Raw Materials, Intermediate Products, Final Products, and Waste Products Generated During Manufacture and Use:

- Zinc

### Photoelectric Cells

Raw Materials, Intermediate Products, Final Products, and Waste Products Generated During Manufacture and Use:

- Thallium

### Photoengraving

General Types of Associated Materials:

- Acids
- Inks
- Solvents

Raw Materials, Intermediate Products, Final Products, and Waste Product
Generated During Manufacture and Use:

- Chromium
- Hydrogen sulfide
- Mercury
- Nitrogen
- Phosphoric acid
- Potassium hydroxide

Other Associated Materials:

- Ammonium compounds
- Hydrogen chloride
- Methyl alcohol

### Photographic Chemicals

General Types of Associated Materials:

- Acids
- Alkalis
- Turpentine

Raw Materials, Intermediate Products, Final Products, and Waste Product
Generated During Manufacture and Use:

- Acetaldehyde
- Acetic acid
- Amyl alcohol
- Aniline
- Benzyl chloride
- Boron
- Cadmium
- Chromates
- Chromium
- Dibromoethane
- Dichloroethane
- Dinitrophenol

Raw Materials, Intermediate Products, Final Products, and Waste Products
Generated During Manufacture and Use (cont.):

- Hydrazine and derivatives
- Hydroquinones
- Mercury
- Methylaminoethanol
- Methyl para-aminophenol sulfate
- n-Butylamine
- Para-aminophenol
- Paraformaldehyde
- Paraphenylenediamines
- Picric acid
- Pyrogallic acids
- Selenium
- Sodium hydroxide
- Sodium hypochlorite
- Sodium sulfide
- Uranium
- Vanadium

Other Associated Materials:

- Hydrogen chloride

## Photographic Films

General Types of Associated Materials:

- Acetates

Raw Materials, Intermediate Products, Final Products, and Waste Products
Generated During Manufacture and Use:

- Formaldehyde
- Ketones
- Methyl alcohol
- n-Butyl alcohol
- Quinone
- Silver

Other Associated Materials:

- Acetic anhydride

## Photographic Paper

Raw Materials, Intermediate Products, Final Products, and Waste Products Generated During Manufacture and Use:

- Silver

## Photographic Plates

Raw Materials, Intermediate Products, Final Products, and Waste Products Generated During Manufacture and Use:

- Silver

## Phthaleins

Raw Materials, Intermediate Products, Final Products, and Waste Products Generated During Manufacture and Use:

- Phthalic anhydride

## Phthalic Acid

Raw Materials, Intermediate Products, Final Products, and Waste Products Generated During Manufacture and Use:

- Nitrogen
- Phthalic anhydride

## Phthalic Anhydride

Raw Materials, Intermediate Products, Final Products, and Waste Products Generated During Manufacture and Use:

- Xylenes

## Phthalic Compounds

Raw Materials, Intermediate Products, Final Products, and Waste Products Generated During Manufacture and Use:

- Naphthalene

## Pigments

Raw Materials, Intermediate Products, Final Products, and Waste Products Generated During Manufacture and Use:

- Antimony
- Arsenic
- Bismuth
- Cadmium
- Copper
- Dichlorobenzidine
- Graphite
- Lead
- Magnesium
- Mercury
- Molybdenum
- Thallium
- Tin
- Titanium
- Tricresyl phosphates
- Zirconium

Other Associated Materials:

- Hydrogen chloride
- Ketones

## Pine Oil

General Types of Associated Materials:

- Turpentine

Raw Materials, Intermediate Products, Final Products, and Waste Products Generated During Manufacture and Use:

- Polynuclear aromatic hydrocarbons

## Pipe Fittings

General Types of Associated Materials:

- Lubricants
- Metals

Raw Materials, Intermediate Products, Final Products, and Waste Products Generated During Manufacture and Use:

- Chromium
- Cyanides

Other Associated Materials:

- Benzene
- Chloroform
- Perchloroethene
- Toluene
- Trichloroethene

## Pipelines

General Types of Associated Materials:

- Brine
- Crude oil
- Fluxes
- Gasoline
- Petroleum fuels
- Petroleum hydrocarbons
- Soaps

## Piping

General Types of Associated Materials:

- Lubricants
- Metals
- Solder

Raw Materials, Intermediate Products, Final Products, and Waste Products Generated During Manufacture and Use:

- Lead
- Silver

## Pitch Processing

General Types of Associated Materials:

- Pitch
- Solvents
- Tar and derivatives

Raw Materials, Intermediate Products, Final Products, and Waste Product Generated During Manufacture and Use:

- Creosote
- Cresol
- Phenols
- Polynuclear aromatic hydrocarbons

## Pival

General Types of Associated Materials:

- Rodenticides

Raw Materials, Intermediate Products, Final Products, and Waste Product Generated During Manufacture and Use:

- Coumarin
- Indandiones

## Pivalyn

General Types of Associated Materials:

- Rodenticides

Raw Materials, Intermediate Products, Final Products, and Waste Product Generated During Manufacture and Use:

- Coumarin
- Indandiones

## Plaster

Raw Materials, Intermediate Products, Final Products, and Waste Products Generated During Manufacture and Use:

- Asbestos
- Calcium oxide

## Plastics

General Types of Associated Materials:

- Accelerators
- Acids
- Activators
- Adhesives and removers
- Alkalis
- Antioxidants
- Diesel fuel
- Heating oil
- Oils
- Plasticizers
- Resins
- Retarders
- Soaps
- Solvents
- Tar and derivatives
- Turpentine

Raw Materials, Intermediate Products, Final Products, and Waste Products Generated During Manufacture and Use:

- Acetic acid
- Acrolein
- Acrylonitrile
- Aluminum
- Aluminum oxide
- Aniline
- Antimony

Raw Materials, Intermediate Products, Final Products, and Waste Product Generated During Manufacture and Use (cont.):

- Benzidine
- Benzyl chloride
- beta-Propiolactone
- Boron hydrides
- Butadiene
- Cadmium
- Chlorine
- Chloroform
- Chloroprene dimers
- Chromium
- Cresol
- Curene MOCA
- Dichlorobenzidine
- Dichloroethane
- Ethylene glycol ether
- Ethylhexylphthalate
- Formaldehyde
- Furfural
- Hexamethylenetetramine
- Isopropyl alcohol
- Ketones
- Lead
- Methyl alcohol
- n-Butyl alcohol
- Nickel
- Phenols
- Phthalates
- Phthalic anhydride
- Polynuclear aromatic hydrocarbons
- Propylene dichloride
- Sodium chloride
- Sodium hydroxide
- Tin
- Titanium
- Tricresyl phosphates
- Zinc

Other Associated Materials:

- Acetic anhydride
- Amyl alcohol
- Benzene
- Benzol
- Dichloroethane
- Dichloroethene
- Ethyl alcohol
- Ethyl benzene
- Hydrogen chloride
- Hydrogen fluoride
- Methylene chloride
- Nitroparaffins
- Nitrosodimethylamine
- Perchloroethene
- Toluene
- Trichloroethane
- Trichloroethene
- Xylenes

## Plastic Molding

General Types of Associated Materials:

- Plastics
- Solvents

Other Associated Materials:

- Cyclohexane
- Cyclohexene
- Cyclopropane
- Methylcyclohexene

## Plastic, ABS

Raw Materials, Intermediate Products, Final Products, and Waste Products Generated During Manufacture and Use:

- Styrene

## Plasticizers

Raw Materials, Intermediate Products, Final Products, and Waste Products Generated During Manufacture and Use:

- Acrolein
- Allyl alcohol
- Antimony
- Bis(2-ethylhexyl) phthalate
- Chlorinated naphthalenes
- Chlorodiphenyls
- Ethanolamines
- n-Butyl alcohol
- Phosphorus
- Tricresyl phosphates

Other Associated Materials:

- Ethyl alcohol

## Plumbing Supplies

General Types of Associated Materials:

- Acids
- Adhesives and removers
- Caulking compounds
- Fluxes
- Solders
- Solvents

Raw Materials, Intermediate Products, Final Products, and Waste Products Generated During Manufacture and Use:

- Arsine
- Lead
- Silver
- Zinc

## Plutonium Production

Raw Materials, Intermediate Products, Final Products, and Waste Products Generated During Manufacture and Use:

- Uranium

## PMP

General Types of Associated Materials:

- Rodenticides

Raw Materials, Intermediate Products, Final Products, and Waste Products Generated During Manufacture and Use:

- Coumarin
- Indandiones

## Polyethylene Terephthalates

General Types of Associated Materials:

- Petroleum hydrocarbons
- Plastics

Raw Materials, Intermediate Products, Final Products, and Waste Products Generated During Manufacture and Use:

- Xylenes

## Polishes

General Types of Associated Materials:

- Abrasives
- Acetates
- Paraffin

Raw Materials, Intermediate Products, Final Products, and Waste Products Generated During Manufacture and Use:

- Ethanolamines
- Graphite
- Hydrogen cyanide
- Propyl alcohol

Other Associated Materials:

- Carbon disulfide
- Chloroform
- Dioxane
- Ethyl alcohol
- Nitrobenzene

## Polyacrylic Fibers

Raw Materials, Intermediate Products, Final Products, and Waste Products Generated During Manufacture and Use:

- Tricresyl phosphates

Other Associated Materials:

- Dimethylformamide

### Polyester Resins

Raw Materials, Intermediate Products, Final Products, and Waste Products Generated During Manufacture and Use:

- Ethyl benzene
- Phthalic anhydride

### Polyethyleneimines

Raw Materials, Intermediate Products, Final Products, and Waste Products Generated During Manufacture and Use:

- Ethyleneimine

### Polyglycol

Raw Materials, Intermediate Products, Final Products, and Waste Products Generated During Manufacture and Use:

- Ethylene oxide

### Polymers

Raw Materials, Intermediate Products, Final Products, and Waste Products Generated During Manufacture and Use:

- Chlorine
- Furfural

Other Associated Materials:

- bis(chloromethyl) Ether
- Nitroparaffins

## Polynuclear Aromatic Hydrocarbons

Raw Materials, Intermediate Products, Final Products, and Waste Products Generated During Manufacture and Use:

- Acenaphthene
- Acenapthylene
- Anthracene
- Benzo(a)anthracene
- Benzo(a)pyrene
- Benzo(b)fluoranthene
- Benzo(ghi)perylene
- Benzo(k)fluoranthene
- Chrysene
- Dibenzo(a,h)anthracene
- Fluoranthene
- Fluorene
- Ideno(1,2,3-cd)pyrene
- Naphthalene
- Phenanthrene
- Pyrene

Other Associated Materials:

- Benzene
- Ethyl benzene
- Toluene
- Xylenes

## Polyoxirane

Raw Materials, Intermediate Products, Final Products, and Waste Products Generated During Manufacture and Use:

- Ethylene oxide

### Polyphosphates

Raw Materials, Intermediate Products, Final Products, and Waste Products Generated During Manufacture and Use:

- Phosphoric acid

### Polystyrene

Raw Materials, Intermediate Products, Final Products, and Waste Products Generated During Manufacture and Use:

- Ethyl benzene
- Styrene
- Tricresyl phosphates

### Polyurethanes

General Types of Associated Materials:

- Resins
- Solvents

Raw Materials, Intermediate Products, Final Products, and Waste Products Generated During Manufacture and Use:

- Dichlorobenzidine
- Isocyanates
- MOCA

## Polyvinyl Chloride

Raw Materials, Intermediate Products, Final Products, and Waste Products Generated During Manufacture and Use:

- Cadmium
- Dichloroethane
- Vinyl chloride

## Porcelain

Raw Materials, Intermediate Products, Final Products, and Waste Products Generated During Manufacture and Use:

- Boron
- Clays
- Cobalt
- Silver
- Tin

## Portland Cement

Raw Materials, Intermediate Products, Final Products, and Waste Products Generated During Manufacture and Use:

- Calcium oxide
- Chromates
- Silica

## Pottery

Raw Materials, Intermediate Products, Final Products, and Waste Products Generated During Manufacture and Use:

- Aluminum
- Antimony
- Boron
- Fluorides

## Poultry

General Types of Associated Materials:

- Farming
- Feeds
- Sewage

Raw Materials, Intermediate Products, Final Products, and Waste Products Generated During Manufacture and Use:

- Bacteria
- Nitrates
- Viruses

## Preservatives

Raw Materials, Intermediate Products, Final Products, and Waste Products Generated During Manufacture and Use:

- Carbon disulfide
- Ethyl alcohol
- Isopropyl alcohol
- Mercury

## Printing

General Types of Associated Materials:

- Alkalis
- Glues
- Gums
- Heavy metals
- Inks
- Lubricants
- Petroleum hydrocarbons
- Resins
- Solvents
- Varnishes

Raw Materials, Intermediate Products, Final Products, and Waste Products Generated During Manufacture and Use:

- Aniline dyes
- Antimony
- Arsenic
- Boron
- Chromates
- Chromium
- Copper
- Hydroquinones
- Lead
- Methylene chloride
- Molybdenum
- Oxalic acid
- Picric acid
- Potassium hydroxide
- Silver
- Toluene
- Vanadium
- Xylenes
- Zinc

Other Associated Materials:

- Benzene
- Dioxane
- Ethylene glycol ether
- Tetrachloroethene
- Toluene
- Trichloroethene

## Procaine

Raw Materials, Intermediate Products, Final Products, and Waste Products
Generated During Manufacture and Use:

- Ethylene chlorohydrin

## Proflavine

Raw Materials, Intermediate Products, Final Products, and Waste Products
Generated During Manufacture and Use:

- Acridine

## Propellants

Raw Materials, Intermediate Products, Final Products, and Waste Products
Generated During Manufacture and Use:

- Alcohols
- Carbon tetrachloride

Other Associated Materials:

- Trichloroethane

## Propyl Alcohol

- See Isopropyl Alcohol
- See n-Propyl Alcohol

## Protein Fibers

Raw Materials, Intermediate Products, Final Products, and Waste Products Generated During Manufacture and Use:

- Quinone

## Protein Processing

Raw Materials, Intermediate Products, Final Products, and Waste Products Generated During Manufacture and Use:

- Sulfur dioxide

## Prothiophos

General Types of Associated Materials:

- Organophosphorus insecticides

Raw Materials, Intermediate Products, Final Products, and Waste Products Generated During Manufacture and Use:

- See Insecticides

## Prussic Acid

Raw Materials, Intermediate Products, Final Products, and Waste Products
Generated During Manufacture and Use:

- Hydrogen cyanide

## Pumps

General Types of Associated Materials:

- Lubricants
- Petroleum fuels
- Solvents

Other Associated Materials:

- Benzene
- Dichloroethane
- Dichloroethene
- Ethyl benzene
- Perchloroethene
- Polynuclear aromatic hydrocarbons
- Toluene
- Trichloroethene
- Xylenes

## Putty

General Types of Associated Materials:

- Plastics
- Resins
- Solvents

Raw Materials, Intermediate Products, Final Products, and Waste Products Generated During Manufacture and Use:

- Carbon disulfide

Other Associated Materials:

- Benzene

## Pyridine

Raw Materials, Intermediate Products, Final Products, and Waste Products Generated During Manufacture and Use:

- Acetaldehyde

## Pyromucic Acid

Raw Materials, Intermediate Products, Final Products, and Waste Products Generated During Manufacture and Use:

- Furfural

## Pyrotechnics

General Types of Associated Materials:

- Explosives
- Incendiaries

Raw Materials, Intermediate Products, Final Products, and Waste Products Generated During Manufacture and Use:

- Aluminum
- Antimony
- Phosphorus
- Titanium

## Quartz Crystal Oscillators

Raw Materials, Intermediate Products, Final Products, and Waste Products Generated During Manufacture and Use:

- Xylenes

## Quinol

Raw Materials, Intermediate Products, Final Products, and Waste Products Generated During Manufacture and Use:

- Hydroquinone

## Radar

Raw Materials, Intermediate Products, Final Products, and Waste Products Generated During Manufacture and Use:

- MOCA

## Radium Salts

Raw Materials, Intermediate Products, Final Products, and Waste Products Generated During Manufacture and Use:

- Uranium

## Railroads

General Types of Associated Materials:

- Alkalis
- Antiseptics
- Cutting fluids
- Detergents
- Diesel fuel
- Fungicides
- Heating oil
- Herbicides
- Insecticides
- Lacquers
- Lubricants
- Paint removers
- Paints
- Petroleum fuels
- Pitch
- Solvents
- Tar and derivatives

Raw Materials, Intermediate Products, Final Products, and Waste Products Generated During Manufacture and Use:

- Arsenic
- Chromates
- Coal
- Creosote
- Lead

Other Associated Materials:

- Benzene
- Dichlorobenzene
- Ethyl benzene
- Toluene
- Trichloroethane
- Trichloroethene
- Vinyl chloride
- Xylenes

## Rayon

General Types of Associated Materials:

- Acids
- Alkalis
- Bleaches
- Oils
- Solvents

Raw Materials, Intermediate Products, Final Products, and Waste Products Generated During Manufacture and Use:

- Ammonia
- Calcium bisulfite
- Carbon disulfide
- Chlorine
- Cyanides
- Hydrogen sulfide
- Nitrogen
- Sodium hydroxide
- Sodium sulfide
- Sodium sulfite
- Titanium
- Zirconium

Other Associated Materials:

- Acetic acid
- Carbon disulfide
- Copper
- Ethyl ether

## Rectifiers

Raw Materials, Intermediate Products, Final Products, and Waste Products Generated During Manufacture and Use:

- Germanium
- Graphite
- Mercury
- Selenium
- Tellurium

## Reducing Agents

Raw Materials, Intermediate Products, Final Products, and Waste Products Generated During Manufacture and Use:

- Dibromoethane

## Refractory Materials

Raw Materials, Intermediate Products, Final Products, and Waste Products Generated During Manufacture and Use:

- Aluminum
- Beryllium
- Calcium oxide
- Chromium
- Graphite
- Zirconium

## Refrigerants

Raw Materials, Intermediate Products, Final Products, and Waste Products Generated During Manufacture and Use:

- Acrolein
- Ammonia
- Carbon tetrachloride
- Chlorine
- Dichloroethene
- Ethyl chloride
- Freon
- Sulfur dioxide

## Refrigeration Equipment

General Types of Associated Materials:

- Brine
- Lubricants
- Refrigerants

Raw Materials, Intermediate Products, Final Products, and Waste Products Generated During Manufacture and Use:

- Chromates
- Chromium
- Lead
- Sulfur dioxide
- Tellurium

Other Associated Materials:

- Ammonia
- Ethyl bromide
- Ethyl chloride
- Methyl chloride
- Trichloroethene

## Rendering

Raw Materials, Intermediate Products, Final Products, and Waste Products Generated During Manufacture and Use:

- See Fat Processing

## Resins

General Types of Associated Materials:

- Acetates
- Turpentine

Raw Materials, Intermediate Products, Final Products, and Waste Products Generated During Manufacture and Use:

- Acrolein
- Acrylonitrile
- Allyl alcohol
- Aniline
- Benzyl chloride
- beta-Propiolactone
- Butadiene
- Chlorinated benzenes
- Chlorodiphenyls
- Cresol
- Dichloroethane
- Epichlorohydrin
- Ethyl chloride
- Ethylene glycol ether
- Ethylenediamine

Raw Materials, Intermediate Products, Final Products, and Waste Products Generated During Manufacture and Use (cont.):

- Formaldehyde
- Hexamethylenetetramine
- Isocyanates
- Isopropyl alcohol
- Methyl alcohol
- MOCA
- Phenols
- Phthalic anhydride
- Tetrachloroethane

Other Associated Materials:

- Acetic anhydride
- Carbon disulfide
- Carbon tetrachloride
- Cyclohexane
- Cyclohexene
- Cyclopropane
- Dibromomethane
- Dichloroethene
- Dimethylformamide
- Ethyl benzene
- Ethylene chlorohydrin
- Ethylene glycol
- Methylcyclohexene
- n-Propyl alcohol
- Nitroparaffins
- Propyl alcohol
- Pyridine
- Trichloroethene
- Vinyl chloride

## Resins, Alkyd

General Types of Associated Materials:

- Resins

Raw Materials, Intermediate Products, Final Products, and Waste Products Generated During Manufacture and Use:

- Phthalic anhydride

## Resins, Ion Exchange

General Types of Associated Materials:

- Resins

Raw Materials, Intermediate Products, Final Products, and Waste Products Generated During Manufacture and Use:

- bis(chloromethyl) Ether

## Resins, Isocyanate

General Types of Associated Materials:

- Resins

Raw Materials, Intermediate Products, Final Products, and Waste Products Generated During Manufacture and Use:

- Isocyanates

### Resins, Polyester

General Types of Associated Materials:

- Resins

Raw Materials, Intermediate Products, Final Products, and Waste Products Generated During Manufacture and Use:

- Phthalic anhydride

### Resins, Synthetic

General Types of Associated Materials:

- Alcohols
- Resins

Raw Materials, Intermediate Products, Final Products, and Waste Products Generated During Manufacture and Use:

- Naphthalene
- Phenols
- Polynuclear aromatic hydrocarbons

Other Associated Materials:

- Ethyl alcohol
- n-Propyl alcohol

### Resins, Thermosetting

General Types of Associated Materials:

- Resins

Raw Materials, Intermediate Products, Final Products, and Waste Products Generated During Manufacture and Use:

- Furfural

### Resins, Urea-formaldehyde

General Types of Associated Materials:

- Resins

Raw Materials, Intermediate Products, Final Products, and Waste Products Generated During Manufacture and Use:

- Hexamethylenetetramine

### Resins, Vinyl

General Types of Associated Materials:

- Resins

Raw Materials, Intermediate Products, Final Products, and Waste Products Generated During Manufacture and Use:

- Phthalic anhydride

### Rivanol

Raw Materials, Intermediate Products, Final Products, and Waste Products
Generated During Manufacture and Use:

- Acridine

### Rocket Fuels

General Types of Associated Materials:

- Gasoline
- Kerosene

Raw Materials, Intermediate Products, Final Products, and Waste Products
Generated During Manufacture and Use:

- Aniline
- Boron
- Boron hydrides
- Butadiene
- Carbon disulfide
- Cerium
- Chlorine trifluoride
- Dimethyl hydrazine
- Ethyl oxide
- Ethylene oxide
- Fluorine
- Hydrazine and derivatives
- Hydrogen fluoride
- Hydrogen peroxide
- Nitric acid
- Nitrogen
- Picric acid

## Rodenticides

Raw Materials, Intermediate Products, Final Products, and Waste Products Generated During Manufacture and Use:

- alpha-Naphthylthiouria
- Arsenic
- Coumarins
- Fluorides
- Fluoroacetate
- Indandiones
- Phosphides
- Phosphorus
- Strychnine
- Tetramethylthiuram disulfide
- Thallium
- Thallium sulfate

## Ronnel

General Types of Associated Materials:

- Organophosphorus insecticides

Raw Materials, Intermediate Products, Final Products, and Waste Products Generated During Manufacture and Use:

- See Insecticides

## Roofing Tiles

Raw Materials, Intermediate Products, Final Products, and Waste Products Generated During Manufacture and Use:

- Tin

## Roofing, Asphalt

Raw Materials, Intermediate Products, Final Products, and Waste Products
Generated During Manufacture and Use:

- Asbestos
- Polynuclear aromatic hydrocarbons

Other Associated Materials:

- Benzene
- Ethyl benzene
- Toluene
- Xylenes

## Rope Making

General Types of Associated Materials:

- Alkalis
- Bleaches
- Dyes
- Oils
- Pitch
- Soaps
- Tar and derivatives

## Rosins

General Types of Associated Materials:

- Acetates

Other Associated Materials:

- Ethylene chlorohydrin
- n-Propyl alcohol

## Rubber Accelerators

Raw Materials, Intermediate Products, Final Products, and Waste Products Generated During Manufacture and Use:

- Ethyl alcohol
- Methyl alcohol
- Tetramethylthiuram disulfide

## Rubber Cements

General Types of Associated Materials:

- Alcohols
- Ketones
- Solvents

Raw Materials, Intermediate Products, Final Products, and Waste Products Generated During Manufacture and Use:

- Carbon disulfide
- n-Butyl alcohol

Other Associated Materials:

- Toluene
- Trichloroethene

## Rubber Manufacturing and Processing

General Types of Associated Materials:

- Accelerators
- Acids
- Activators
- Adhesives and removers
- Alkalis

General Types of Associated Materials (cont.):

- Antioxidants
- Diesel fuel
- Heating oil
- Naphtha
- Oils
- Plasticizers
- Resins
- Retarders
- Soaps
- Solvents
- Tar and derivatives
- Turpentine

Raw Materials, Intermediate Products, Final Products, and Waste Products Generated During Manufacture and Use:

- Acetaldehyde
- Acetic acid
- Acrolein
- Acrylonitrile
- alpha-Naphthylamine
- Aluminum
- Aluminum oxide
- Aminodiphenyl
- Aniline
- Antimony
- Benzidine
- Benzyl chloride
- beta-Naphthylamine
- Boron hydrides
- Chlorinated naphthalenes
- Chlorine
- Chlorodiphenyls
- Chloroprene
- Chloroprene dimers
- Chromium
- Cobalt
- Curene MOCA
- Dichloroethane

Raw Materials, Intermediate Products, Final Products, and Waste Products Generated During Manufacture and Use (cont.):

- Dichloroethene
- Ethyl benzene
- Ethylenediamine
- Ethylhexylphthalate
- Formaldehyde
- Formic acid
- Hexamethylenetetramine
- Isocyanates
- Ketones
- Lead
- n-Butylamine
- Oxalic acid
- Phenols
- Phosphorus
- Polynuclear aromatic hydrocarbons
- Propylene dichloride
- Sodium chloride
- Sodium hydroxide
- Styrene
- Sulfur chloride
- Tetrachloroethene
- Tetramethylthiuram disulfide
- Tin
- Titanium
- Tricresyl phosphates
- Zinc

Other Associated Materials:

- Amyl alcohol
- Benzene
- Benzol
- Carbon tetrachloride
- Cyclohexane
- Cyclohexene
- Cyclopropane
- Dichloroethane
- Dichloroethene

Other Associated Materials (cont.):

- Ethyl alcohol
- Ethyl benzene
- Ethyl ether
- Hydrogen chloride
- Methylcyclohexene
- Methylene chloride
- Nitroparaffins
- Perchloroethene
- Pyridine
- Toluene
- Trichloroethane
- Trichloroethene
- Vinyl chloride
- Xylenes

## Rubber, Natural

Raw Materials, Intermediate Products, Final Products, and Waste Products
Generated During Manufacture and Use:

- Ethylenediamine
- Formic acid
- Furfural
- Lead
- Phosphoric acid
- Selenium
- Sulfur chloride
- Tellurium

Other Associated Materials:

- Carbon disulfide

## Rum

General Types of Associated Materials:

- Alcohols

Raw Materials, Intermediate Products, Final Products, and Waste Products
Generated During Manufacture and Use:

- Formates

## Rust Inhibitors

Raw Materials, Intermediate Products, Final Products, and Waste Products
Generated During Manufacture and Use:

- Phosphoric acid

## Saccharin

Raw Materials, Intermediate Products, Final Products, and Waste Products
Generated During Manufacture and Use:

- Phosphorus
- Toluene

## Salt Storage

General Types of Associated Materials:

- Brines

.

Raw Materials, Intermediate Products, Final Products, and Waste Products
Generated During Manufacture and Use:

- Magnesium
- Sodium chloride

## Sanitary Services

- See Sewage

## Scrap Metal Reclamation

General Types of Associated Materials:

- Grease
- Lubricants
- Petroleum fuels
- Solvents

Raw Materials, Intermediate Products, Final Products, and Waste Products
Generated During Manufacture and Use:

- Cadmium
- Chromium
- Copper
- Lead
- Nickel
- Polybrominated biphenyls
- Polychlorinated biphenyls
- Polynuclear aromatic hydrocarbons

Other Associated Materials:

- Benzene
- Ethyl benzene
- Perchloroethene
- Toluene

Other Associated Materials (cont.):

- Trichloroethane
- Trichloroethene
- Xylenes

## Sculpting

General Types of Associated Materials:

- Plaster of paris
- Polishes

## Seed Treatment

Raw Materials, Intermediate Products, Final Products, and Waste Products Generated During Manufacture and Use:

- Chlorinated benzenes
- Mercury
- Mercury, alkyl
- Tetramethylthiuram disulfide

Other Associated Materials:

- Ethylene chlorohydrin

## Selenium

Other Associated Materials:

- Carbon disulfide

## Semiconductors

Raw Materials, Intermediate Products, Final Products, and Waste Products Generated During Manufacture and Use:

- Arsenic
- Cadmium
- Germanium
- Phosphorus
- Selenium
- Tellurium

## Service Station

- See Automotive Service Stations

## Sevin

General Types of Associated Materials:

- Carbamate insecticides

Raw Materials, Intermediate Products, Final Products, and Waste Products Generated During Manufacture and Use:

- See Insecticides

## Shampoos

Raw Materials, Intermediate Products, Final Products, and Waste Products Generated During Manufacture and Use:

- Ethanolamines

## Shaving Preparations

Raw Materials, Intermediate Products, Final Products, and Waste Products
Generated During Manufacture and Use:

- Ethanolamines

## Shellac

Raw Materials, Intermediate Products, Final Products, and Waste Products
Generated During Manufacture and Use:

- Ethylenediamine
- n-Butyl alcohol

Other Associated Materials:

- Ethyl alcohol
- n-Propyl alcohol
- Nitroparaffins

## Shipyards

General Types of Associated Materials:

- Fungicides
- Paint removers
- Paint thinners
- Paints
- Resins
- Solvents
- Tar and derivatives
- Wood preservatives

Raw Materials, Intermediate Products, Final Products, and Waste Products
Generated During Manufacture and Use:

- Chlorinated diphenyls
- Chromates
- Naphthalenes, chlorinated
- Polychlorinated biphenyls

## Shoe Creams

General Types of Associated Materials:

- Oils
- Polishes
- Solvents
- Waxes

Raw Materials, Intermediate Products, Final Products, and Waste Products
Generated During Manufacture and Use:

- Dioxane

## Shoe Manufacturing and Repair

General Types of Associated Materials:

- Adhesives and removers
- Fungicides
- Naphtha
- Polishes
- Resins
- Rubber
- Solvents
- Tanning agents
- Waxes

Raw Materials, Intermediate Products, Final Products, and Waste Products Generated During Manufacture and Use:

- Aniline
- Aniline dyes
- Ketones
- Trichloroethene

Other Associated Materials:

- Ammonia
- Amyl acetate
- Amyl alcohol
- Benzene
- Benzol
- Hexane

## Shrinkproofing

Raw Materials, Intermediate Products, Final Products, and Waste Products Generated During Manufacture and Use:

- Ethyleneimine

## Signs and Advertising

Raw Materials, Intermediate Products, Final Products, and Waste Products Generated During Manufacture and Use:

- Caustic soda

Other Associated Materials:

- Ketones
- Toluene

## Silk

Raw Materials, Intermediate Products, Final Products, and Waste Products Generated During Manufacture and Use:

- Hydrogen fluoride
- Hydrogen sulfide

## Silk, Bleaching

Raw Materials, Intermediate Products, Final Products, and Waste Products Generated During Manufacture and Use:

- Bromine

## Silk, Synthetic

General Types of Associated Materials:

- Acetates

Raw Materials, Intermediate Products, Final Products, and Waste Products Generated During Manufacture and Use:

- Chloroform
- Formaldehyde
- Ketones
- Mercury
- Tetrachloroethane

Other Associated Materials:

- Hydrogen chloride

## Silver

Raw Materials, Intermediate Products, Final Products, and Waste Products Generated During Manufacture and Use:

- Arsenic
- Chlorine
- Hydrogen cyanide
- Mercury

Other Associated Materials:

- Fluorides

## Silvex

Raw Materials, Intermediate Products, Final Products, and Waste Products Generated During Manufacture and Use:

- Ethanolamines

## Skin Lotions

Raw Materials, Intermediate Products, Final Products, and Waste Products Generated During Manufacture and Use:

- Isopropyl alcohol
- Titanium

## Slaughterhouses

- See Butchers

### Smelting

* See Ore Smelting and Processing

### Smokeless Powder

General Types of Associated Materials:

* Acetates
* Explosives

Raw Materials, Intermediate Products, Final Products, and Waste Products Generated During Manufacture and Use:

* Ethyl ether
* Ketones
* Naphthalene
* Nitrocellulose

### Soap

General Types of Associated Materials:

* Alkalis
* Bacteriostats
* Detergents
* Perfumes

Raw Materials, Intermediate Products, Final Products, and Waste Products Generated During Manufacture and Use:

* Boron
* Chlorinated lime
* Ethanolamines
* Hydrogen sulfide
* Isopropyl alcohol
* Nickel

Raw Materials, Intermediate Products, Final Products, and Waste Products Generated During Manufacture and Use (cont.):

- Phenols
- Potassium hydroxide
- Propyl alcohol
- Sodium hydroxide
- Tetrachloroethene
- Tetramethylthiuram disulfide
- Trichloroethene

Other Associated Materials:

- Dichloroethyl ether
- Hydrogen chloride

## Sodium Naphthionate

Raw Materials, Intermediate Products, Final Products, and Waste Products Generated During Manufacture and Use:

- alpha-Naphthylamine

## Sodium Sulfite

Raw Materials, Intermediate Products, Final Products, and Waste Products Generated During Manufacture and Use:

- Sulfur dioxide

## Softeners

Raw Materials, Intermediate Products, Final Products, and Waste Products Generated During Manufacture and Use:

- Ethanolamines

## Soil Disinfectants

General Types of Associated Materials:

- Fumigants

Raw Materials, Intermediate Products, Final Products, and Waste Products Generated During Manufacture and Use:

- Carbon disulfide

## Solder

General Types of Associated Materials:

- Acids
- Fluxes

Raw Materials, Intermediate Products, Final Products, and Waste Products Generated During Manufacture and Use:

- Antimony
- Arsine
- Bismuth
- Boron
- Cadmium
- Copper
- Cyanides
- Hydrazine and derivatives
- Lead
- Rosin
- Silver
- Stibine
- Tin
- Zinc

Solvents

General Types of Associated Materials:

- Gasoline
- Ketones
- Naphtha
- Turpentine

Raw Materials, Intermediate Products, Final Products, and Waste Products Generated During Manufacture and Use:

- Acetonitrile
- Carbon tetrachloride
- Chlorinated benzenes
- Chlorinated naphthalenes
- Chloroform
- Dichloroethane
- Dichloroethene
- Dimethyl sulfate
- Dimethylacetamide
- Dimethylformamide
- Epichlorohydrin
- Ethyl alcohol
- Ethyl benzene
- Ethylenediamine
- Methylene chloride
- n-Heptane
- n-Propyl alcohol
- Nitrobenzene
- Nitrosodimethylamine
- Propylene dichloride
- Pyridine
- Tetrachloroethane
- Tetrachloroethene
- Toluene
- Trichloroethane
- Trichloroethene
- Tricresyl phosphates
- Vinyl chloride
- Xylenes

## Soya Bean Processing

Raw Materials, Intermediate Products, Final Products, and Waste Products Generated During Manufacture and Use:

- Trichloroethane

## Spark Plugs

General Types of Associated Materials:

- Ceramics
- Metals
- Solvents

Raw Materials, Intermediate Products, Final Products, and Waste Products Generated During Manufacture and Use:

- Beryllium
- Molybdenum
- Platinum

## Stain Removers

General Types of Associated Materials:

- Solvents

Raw Materials, Intermediate Products, Final Products, and Waste Products Generated During Manufacture and Use:

- Acetic acid
- Methylene chloride
- Propylene dichloride
- Trichloroethane

## Stains, Wood

General Types of Associated Materials:

- Naphtha
- Pigments
- Solvents
- Turpentine

Raw Materials, Intermediate Products, Final Products, and Waste Products Generated During Manufacture and Use:

- Cresol
- n-Butyl alcohol

Other Associated Materials:

- Ethyl alcohol
- Ethylene glycol ether
- Ketones
- Methyl alcohol
- Nitroparaffins
- Propyl alcohol

## Steel Alloys

General Types of Associated Materials:

- Foundries
- Metals

Raw Materials, Intermediate Products, Final Products, and Waste Products Generated During Manufacture and Use:

- Bismuth
- Fluorides
- Graphite
- Magnesium
- Molybdenum

Raw Materials, Intermediate Products, Final Products, and Waste Products Generated During Manufacture and Use (cont.):

- Nickel
- Silver
- Titanium
- Uranium
- Vanadium

### Steel Blueing

Raw Materials, Intermediate Products, Final Products, and Waste Products Generated During Manufacture and Use:

- Antimony

### Steel Etching

Raw Materials, Intermediate Products, Final Products, and Waste Products Generated During Manufacture and Use:

- Picric acid

### Steel Hardening

- See Case Hardening

### Steel Refining

Raw Materials, Intermediate Products, Final Products, and Waste Products Generated During Manufacture and Use:

- Calcium cyanamide

## Sterilizing Agents

Raw Materials, Intermediate Products, Final Products, and Waste Products Generated During Manufacture and Use:

- beta-Propiolactone

## Stiffening

Raw Materials, Intermediate Products, Final Products, and Waste Products Generated During Manufacture and Use:

- Ethyleneimine

## Stirophos

General Types of Associated Materials:

- Organophosphorus insecticides

Raw Materials, Intermediate Products, Final Products, and Waste Products Generated During Manufacture and Use:

- See Insecticides

## Stockyards

General Types of Associated Materials:

- Fungicides
- Insecticides

Raw Materials, Intermediate Products, Final Products, and Waste Products Generated During Manufacture and Use:

- Bacteria
- Portland cement

## Stoddard Solvent

General Types of Associated Materials:

- Petroleum distillates
- Solvents

Raw Materials, Intermediate Products, Final Products, and Waste Products Generated During Manufacture and Use:

- Polynuclear aromatic hydrocarbons

## Stone Cleaning

Raw Materials, Intermediate Products, Final Products, and Waste Products Generated During Manufacture and Use:

- Hydrogen fluoride
- Muriatic acid

## Stone Coating

Raw Materials, Intermediate Products, Final Products, and Waste Products Generated During Manufacture and Use:

- Hydroquinone

## Storage Batteries

Raw Materials, Intermediate Products, Final Products, and Waste Products
Generated During Manufacture and Use:

- Antimony
- Cadmium
- Chlorinated naphthalenes
- Lead
- Lithium
- Nickel
- Stibine
- Sulfuric acid

## Straw

Raw Materials, Intermediate Products, Final Products, and Waste Products
Generated During Manufacture and Use:

- Chlorinated lime

## Stucco

Raw Materials, Intermediate Products, Final Products, and Waste Products
Generated During Manufacture and Use:

- Calcium oxide

## Styrene

General Types of Associated Materials:

- Coal
- Fiber glass
- Resins

Raw Materials, Intermediate Products, Final Products, and Waste Products Generated During Manufacture and Use:

- Benzene
- Ethyl benzene
- Hydroquinone

## Submarines

Raw Materials, Intermediate Products, Final Products, and Waste Products Generated During Manufacture and Use:

- Arsine

## Sugar Refining

General Types of Associated Materials:

- Acids

Raw Materials, Intermediate Products, Final Products, and Waste Products Generated During Manufacture and Use:

- Calcium oxide
- Hydrogen sulfide
- Phosphoric acid
- Sulfur chloride
- Sulfur dioxide

## Sulfa Drugs

Other Associated Materials:

- Pyridine

## Sulfonated Naphthylamines

Raw Materials, Intermediate Products, Final Products, and Waste Products Generated During Manufacture and Use:

- alpha-Naphthylamine

## Sulfonic Acids

Raw Materials, Intermediate Products, Final Products, and Waste Products Generated During Manufacture and Use:

- Isopropyl alcohol

## Sulfonic Compounds

Raw Materials, Intermediate Products, Final Products, and Waste Products Generated During Manufacture and Use:

- Naphthalene

## Sulfur

Raw Materials, Intermediate Products, Final Products, and Waste Products Generated During Manufacture and Use:

- Ethyl chloride

Other Associated Materials:

- Carbon disulfide

### Sulfuric Acid

Raw Materials, Intermediate Products, Final Products, and Waste Products
Generated During Manufacture and Use:

- Ammonia
- Hydrogen sulfide
- Nitrogen
- Selenium
- Sulfur dioxide

### Suntan Products

Raw Materials, Intermediate Products, Final Products, and Waste Products
Generated During Manufacture and Use:

- Tetramethylthiuram disulfide
- Zinc

### Surfactants

Raw Materials, Intermediate Products, Final Products, and Waste Products
Generated During Manufacture and Use:

- Ethanolamines
- Ethylene oxide
- Ethylenediamine
- Phosphorus

### Surgical Instruments

Raw Materials, Intermediate Products, Final Products, and Waste Products
Generated During Manufacture and Use:

- Nickel
- Titanium

## Swimming Pools

General Types of Associated Materials:

- Oxidizers

Raw Materials, Intermediate Products, Final Products, and Waste Products
Generated During Manufacture and Use:

- Chlorine

## Tank Linings

Raw Materials, Intermediate Products, Final Products, and Waste Products
Generated During Manufacture and Use:

- Lead
- Resins

## Tanning

General Types of Associated Materials:

- Brine
- Dyes
- Oils
- Pancreatic extracts
- Petroleum fuels
- Solvents

Raw Materials, Intermediate Products, Final Products, and Waste Products
Generated During Manufacture and Use:

- Alum
- Ammonia
- Arsenic
- Bacteria

Raw Materials, Intermediate Products, Final Products, and Waste Products
Generated During Manufacture and Use (cont.):

- Benzyl chloride
- Boron
- Cadmium
- Calcium oxide
- Chromium
- Cyanides
- Dimethylamine
- Formaldehyde
- Hydrogen sulfide
- Lead
- Mercury
- Molybdenum
- n-Butylamine
- Naphthalene
- Oxalic acid
- Phenols
- Quinone
- Sodium hydroxide
- Sodium sulfide
- Sulfur dioxide
- Sulfuric acid
- Tannin
- Zinc

Other Associated Materials:

- Acetic acid
- Ammonium compounds
- Benzene
- Benzol
- Calcium hydrosulfide
- Ethyl benzene
- Hydrogen chloride
- Perchloroethene
- Toluene
- Xylenes

## Tantalum Ore

Other Associated Materials:

- Hydrogen chloride

## Tar Processing and Use

General Types of Associated Materials:

- Pitch
- Solvents
- Tar and derivatives

Other Associated Materials:

- Dichloroethyl ether
- Tetrachloroethene
- Trichloroethene

## Tartar Emetic

Raw Materials, Intermediate Products, Final Products, and Waste Products Generated During Manufacture and Use:

- Antimony

## Tartaric Acid

Raw Materials, Intermediate Products, Final Products, and Waste Products Generated During Manufacture and Use:

- Oxalic acid

## Taxidermy

General Types of Associated Materials:

- Bacteria
- Metals
- Solvents

Raw Materials, Intermediate Products, Final Products, and Waste Products Generated During Manufacture and Use:

- Anthrax bacillus
- Arsenic
- Mercury
- Tannin
- Zinc

Other Associated Materials:

- Alum

## Terephthalic Acid

Raw Materials, Intermediate Products, Final Products, and Waste Products Generated During Manufacture and Use:

- Xylenes

## Termites

Raw Materials, Intermediate Products, Final Products, and Waste Products Generated During Manufacture and Use:

- Dibromomethane

### Tetrachloroethene

Raw Materials, Intermediate Products, Final Products, and Waste Products Generated During Manufacture and Use:

- Tetrachloroethane

### Tetrachlorvinphos

General Types of Associated Materials:

- Organophosphorus insecticides

Raw Materials, Intermediate Products, Final Products, and Waste Products Generated During Manufacture and Use:

- See Insecticides

### Tetraethyl Lead

Raw Materials, Intermediate Products, Final Products, and Waste Products Generated During Manufacture and Use:

- Dibromomethane
- Ethyl chloride
- Lead

### Tetryl

Raw Materials, Intermediate Products, Final Products, and Waste Products Generated During Manufacture and Use:

- Aniline

## Textile Dyeing

General Types of Associated Materials:

- Dyes
- Textiles

Raw Materials, Intermediate Products, Final Products, and Waste Products Generated During Manufacture and Use:

- Ethylene chlorohydrin

## Textile Finishes

General Types of Associated Materials:

- Flameproofing
- Textiles
- Waterproofing

Raw Materials, Intermediate Products, Final Products, and Waste Products Generated During Manufacture and Use:

- Acrylonitrile

## Textile Printing

General Types of Associated Materials:

- Dyes
- Ink
- Pigments
- Textiles

Raw Materials, Intermediate Products, Final Products, and Waste Products Generated During Manufacture and Use:

- Ethylene chlorohydrin

## Textile Resins

General Types of Associated Materials:

- Resins
- Solvents
- Textiles

Raw Materials, Intermediate Products, Final Products, and Waste Products Generated During Manufacture and Use:

- Acrolein

## Textile Soaps

General Types of Associated Materials:

- Alcohols
- Soaps
- Textiles

Raw Materials, Intermediate Products, Final Products, and Waste Products Generated During Manufacture and Use:

- Methyl alcohol

## Textiles

Raw Materials, Intermediate Products, Final Products, and Waste Products Generated During Manufacture and Use:

- Acetic acid
- Acetic anhydride
- Acrolein
- Arsenic
- Benzene
- Boron

Raw Materials, Intermediate Products, Final Products, and Waste Products
Generated During Manufacture and Use (cont.)

- Calcium chloride
- Carbon disulfide
- Cerium
- Chlorine
- Chlorodiphenyls
- Chromium
- Dichlorobenzidine
- Dichloroethane
- Ethanolamines
- Ethylene glycol
- Ethylene glycol ether
- Ethyleneimine
- Formaldehyde
- Formic acid
- Hexamethylenetetramine
- Hydrazine and derivatives
- Isocyanates
- Isopropyl alcohol
- Ketones
- Magnesium
- Mercury
- Methyl alcohol
- n-Butylamine
- Naphthalene
- Nickel
- Nitrogen
- Oxalic acid
- Phenols
- Picric acid
- Quinone
- Selenium
- Sulfur chloride
- Sulfuric acid
- Trichloroethene
- Vanadium
- Zinc
- Zirconium

Other Associated Materials:

- Amyl alcohol
- Dichloroethyl ether
- Hydrogen chloride
- Nitrosodimethylamine

## Textiles, Acrylic

General Types of Associated Materials:

- Acrylics
- Textiles

Raw Materials, Intermediate Products, Final Products, and Waste Products Generated During Manufacture and Use:

- Hydrazine and derivatives

## Textiles, Synthetic

General Types of Associated Materials:

- Resins
- Solvents
- Textiles

Raw Materials, Intermediate Products, Final Products, and Waste Products Generated During Manufacture and Use:

- Xylenes

## Textiles, Vinyl

General Types of Associated Materials:

- Textiles
- Vinyl

Raw Materials, Intermediate Products, Final Products, and Waste Products Generated During Manufacture and Use:

- Hydrazine and derivatives

## Thermal Insulators

Raw Materials, Intermediate Products, Final Products, and Waste Products Generated During Manufacture and Use:

- Titanium

## Thermocoupling Agents

Raw Materials, Intermediate Products, Final Products, and Waste Products Generated During Manufacture and Use:

- Tellurium

## Thermoelectric Devices

Raw Materials, Intermediate Products, Final Products, and Waste Products Generated During Manufacture and Use:

- Tellurium

## Thermometers

Raw Materials, Intermediate Products, Final Products, and Waste Products Generated During Manufacture and Use:

- Mercury
- n-Hexane
- Sulfur dioxide
- Thallium

## Thiodan

General Types of Associated Materials:

- Organochlorine insecticides

Raw Materials, Intermediate Products, Final Products, and Waste Products Generated During Manufacture and Use:

- See Insecticides

## Thiourea Resins

General Types of Associated Materials:

- Resins

Raw Materials, Intermediate Products, Final Products, and Waste Products Generated During Manufacture and Use:

- Formaldehyde

## Thiram

General Types of Associated Materials:

- Carbamate insecticides

Raw Materials, Intermediate Products, Final Products, and Waste Products Generated During Manufacture and Use:

- See Insecticides

## Tin

Raw Materials, Intermediate Products, Final Products, and Waste Products Generated During Manufacture and Use:

- Arsine
- Bismuth
- Chlorine
- Sodium hydroxide

Other Associated Materials:

- Hydrogen chloride

## Tint Rinses

Raw Materials, Intermediate Products, Final Products, and Waste Products Generated During Manufacture and Use:

- Acetic acid

## Titanium Dioxide

Raw Materials, Intermediate Products, Final Products, and Waste Products
Generated During Manufacture and Use:

- Sulfuric acid

## Tobacco

Raw Materials, Intermediate Products, Final Products, and Waste Products
Generated During Manufacture and Use:

- Ethylene glycol
- Formates
- Tetrachloroethane
- Trichloroethene

## Tokuthion

General Types of Associated Materials:

- Organophosphorus insecticides

Raw Materials, Intermediate Products, Final Products, and Waste Products
Generated During Manufacture and Use:

- See Insecticides

## Toluene Diisocyanate

Raw Materials, Intermediate Products, Final Products, and Waste Products
Generated During Manufacture and Use:

- Isocyanate
- Toluene

### Toluene Sulfonates

Raw Materials, Intermediate Products, Final Products, and Waste Products Generated During Manufacture and Use:

- Toluene

### Toluol

Raw Materials, Intermediate Products, Final Products, and Waste Products Generated During Manufacture and Use:

- Toluene

### Tool and Die

- See Metal Working and Machining

### Toxaphene

General Types of Associated Materials:

- Organochlorine insecticides

Raw Materials, Intermediate Products, Final Products, and Waste Products Generated During Manufacture and Use:

- See Insecticides

## Transformers

General Types of Associated Materials:

- Transformer oils

Raw Materials, Intermediate Products, Final Products, and Waste Products Generated During Manufacture and Use:

- Chlorinated naphthalenes
- Chlorodiphenyls
- Polychlorinated biphenyls
- Polynuclear aromatic hydrocarbons

## Transistors

Raw Materials, Intermediate Products, Final Products, and Waste Products Generated During Manufacture and Use:

- Germanium

## Transportation Services

General Types of Associated Materials:

- Lubricants
- Metals
- Paints
- Polychlorinated biphenyls
- Petroleum fuels
- Solvents

## Trichloronate

General Types of Associated Materials:

- Organophosphorus insecticides

Raw Materials, Intermediate Products, Final Products, and Waste Products Generated During Manufacture and Use:

- See Insecticides

## Trucking Facilities and Repair

General Types of Associated Materials:

- Diesel fuel
- Heating oil
- Lubricants
- Petroleum fuels
- Solvents

Raw Materials, Intermediate Products, Final Products, and Waste Products Generated During Manufacture and Use:

- Heavy metals
- Phenols
- Polychlorinated biphenyls

Other Associated Materials:

- Benzene
- Ethyl benzene
- Formaldehyde
- Methylene chloride
- Toluene
- Trichloroethane
- Xylenes

## Tumor-inhibiting Agents

Raw Materials, Intermediate Products, Final Products, and Waste Products Generated During Manufacture and Use:

- Hydroquinone

## Tungsten

Raw Materials, Intermediate Products, Final Products, and Waste Products Generated During Manufacture and Use:

- Molybdenum

## Turpentine

Raw Materials, Intermediate Products, Final Products, and Waste Products Generated During Manufacture and Use:

- Alpha pinene
- Beta pinene
- Camphene
- Monocyclic terpenes
- Terpene alcohols

## Undertakers

- See Embalming

## Upholstery

General Types of Associated Materials:

- Flame retardants
- Glues
- Lacquer thinners
- Lacquers

Raw Materials, Intermediate Products, Final Products, and Waste Products Generated During Manufacture and Use:

- Bacteria

Other Associated Materials:

- Methyl alcohol

## Uranium Processing

Raw Materials, Intermediate Products, Final Products, and Waste Products Generated During Manufacture and Use:

- Fluorine
- Sulfuric acid

## Urea Resins

General Types of Associated Materials:

- Resins

Raw Materials, Intermediate Products, Final Products, and Waste Products Generated During Manufacture and Use:

- Acetaldehyde

## Urea-formaldehyde Resins

General Types of Associated Materials:

- Resins

Raw Materials, Intermediate Products, Final Products, and Waste Products Generated During Manufacture and Use:

- Formaldehyde
- n-Butyl alcohol

## Ureas

Raw Materials, Intermediate Products, Final Products, and Waste Products Generated During Manufacture and Use:

- Formaldehyde

## Urethanes

Raw Materials, Intermediate Products, Final Products, and Waste Products Generated During Manufacture and Use:

- Dichlorobenzidine
- MOCA

## Vacuum Tubes

Raw Materials, Intermediate Products, Final Products, and Waste Products Generated During Manufacture and Use:

- Germanium
- Molybdenum
- Thorium
- Titanium

Other Associated Materials:

- Tetrachloroethene

## Valone

General Types of Associated Materials:

- Rodenticides

Raw Materials, Intermediate Products, Final Products, and Waste Products Generated During Manufacture and Use:

- Coumarin
- Indandiones

## Valve Manufacturing and Repair

General Types of Associated Materials:

- Metals
- Solvents

Raw Materials, Intermediate Products, Final Products, and Waste Products Generated During Manufacture and Use:

- Chromium
- Cyanides

Other Associated Materials:

- Benzene
- Chloroform
- Perchloroethene
- Toluene
- Trichloroethene

## Vapam

General Types of Associated Materials:

- Carbamate insecticides

Raw Materials, Intermediate Products, Final Products, and Waste Products Generated During Manufacture and Use:

- See Insecticides

## Vapona

General Types of Associated Materials:

- Organophosphate insecticides

Raw Materials, Intermediate Products, Final Products, and Waste Products Generated During Manufacture and Use:

- See Insecticides

## Varnish Removers

General Types of Associated Materials:

- Ketones
- Solvents

Raw Materials, Intermediate Products, Final Products, and Waste Products Generated During Manufacture and Use:

- Potassium hydroxide

Other Associated Materials:

- Cyclohexane
- Cyclohexene
- Cyclopropane
- Dichloroethyl ether
- Dioxane
- Methyl alcohol
- Methylcyclohexene
- Nitroparaffins

## Varnish Thinners

General Types of Associated Materials:

- Solvents

Raw Materials, Intermediate Products, Final Products, and Waste Products Generated During Manufacture and Use:

- Methylene chloride
- Polynuclear aromatic hydrocarbons
- Toluene
- Trichloroethane
- Xylenes

## Varnishes

General Types of Associated Materials:

- Acetates
- Naphtha
- Paraffin
- Solvents
- Turpentine

Raw Materials, Intermediate Products, Final Products, and Waste Products
Generated During Manufacture and Use:

- Aniline
- Epichlorohydrin
- Furfural
- Lead
- Manganese
- Nickel
- Oxalic acid
- Potassium hydroxide
- Titanium
- Zinc

Other Associated Materials:

- Amyl alcohol
- Carbon tetrachloride
- Dichloroethane
- Dichloroethyl ether
- Ethyl benzene
- Ethylene glycol ether
- Ketones
- n-Butyl alcohol
- Tetrachloroethane
- Trichloroethene
- Xylenes

## Vegetable Oil Processing

General Types of Associated Materials:

- Alcohols
- Solvents

Raw Materials, Intermediate Products, Final Products, and Waste Products
Generated During Manufacture and Use:

- Acrolein
- n-Propyl alcohol

## Veterinary Facilities

General Types of Associated Materials:

- Deodorants
- Detergents
- Drugs
- Pesticides
- Soaps

Raw Materials, Intermediate Products, Final Products, and Waste Products Generated During Manufacture and Use:

- Bacteria
- Mercury
- Viruses

## Vinegar

Raw Materials, Intermediate Products, Final Products, and Waste Products Generated During Manufacture and Use:

- Acetaldehyde

## Vineyards

Raw Materials, Intermediate Products, Final Products, and Waste Products Generated During Manufacture and Use:

- Sulfur dioxide

## Vinyl

General Types of Associated Materials:

- Ketones
- Plastics
- Solvents

Raw Materials, Intermediate Products, Final Products, and Waste Products Generated During Manufacture and Use:

- Acetic acid
- Nitroparaffins
- Phthalic anhydride
- Tricresyl phosphates

## Vinyl Chloride

Raw Materials, Intermediate Products, Final Products, and Waste Products Generated During Manufacture and Use:

- Dichloroethane
- Trichloroethane

Other Associated Materials:

- Hydrogen chloride

## Vinyl Toluene

Raw Materials, Intermediate Products, Final Products, and Waste Products Generated During Manufacture and Use:

- Toluene

## Viricides

Raw Materials, Intermediate Products, Final Products, and Waste Products Generated During Manufacture and Use:

- beta-Propiolactone

## Viscose Rayon

Raw Materials, Intermediate Products, Final Products, and Waste Products Generated During Manufacture and Use:

- Dichloroethane

## Vitamins

Other Associated Materials:

- Pyridine

## Wallpaper

General Types of Associated Materials:

- Glues
- Paper

Raw Materials, Intermediate Products, Final Products, and Waste Products Generated During Manufacture and Use:

- Copper

## Warfarin

General Types of Associated Materials:

- Rodenticides

Raw Materials, Intermediate Products, Final Products, and Waste Products Generated During Manufacture and Use:

- Coumarin
- Indandiones

## Watch Manufacturing and Repair

- See Jewelry Manufacturing and Repair

## Water Treatment Chemicals

General Types of Associated Materials:

- Bactericides
- Oxidizers

Raw Materials, Intermediate Products, Final Products, and Waste Products Generated During Manufacture and Use:

- Ammonia
- Copper
- Dichloroethane
- Fluorides .
- Hydrazine and derivatives
- Manganese
- Phosphoric acid
- Silver

## Waterproofing

General Types of Associated Materials:

- Oils
- Paraffin
- Pitch
- Resins
- Rubber
- Solvents
- Tar and derivatives
- Waxes

Raw Materials, Intermediate Products, Final Products, and Waste Products Generated During Manufacture and Use:

- Aluminum salts
- Creosote
- Dibromomethane
- Ethyleneimine
- Formaldehyde
- Methyl alcohol

## Waxes

General Types of Associated Materials:

- Naphtha
- Turpentine

Raw Materials, Intermediate Products, Final Products, and Waste Products Generated During Manufacture and Use:

- Ethanolamines
- Ethyl chloride
- Ethylene glycol
- Phosphoric acid
- Polynuclear aromatic hydrocarbons

Other Associated Materials:

- Amyl alcohol
- Carbon disulfide
- Carbon tetrachloride
- Cyclohexane
- Cyclohexene
- Cyclopropane
- Dibromomethane
- Dichloroethene
- Dioxane
- Ethyl ether
- Ethylene chlorohydrin
- Ethylene glycol ether
- Ketones
- Methylcyclohexene
- Methylene chloride
- Nitroparaffins
- Propyl alcohol
- Propylene dichloride
- Tetrachloroethane
- Tetrachloroethene
- Trichloroethene

## Waxes, Synthetic

General Types of Associated Materials:

- Waxes

Raw Materials, Intermediate Products, Final Products, and Waste Products Generated During Manufacture and Use:

- Ethylenediamine

## Wax Paper

General Types of Associated Materials:

- Paper
- Paraffin

## Weatherproofing

Raw Materials, Intermediate Products, Final Products, and Waste Products Generated During Manufacture and Use:

- Boron
- Silicone

## Weed Killers

- See Herbicides

## Welding

General Types of Associated Materials:

- Fluxes

Raw Materials, Intermediate Products, Final Products, and Waste Products Generated During Manufacture and Use:

- Chromium
- Fluorides
- Nitrogen
- Thorium
- Titanium
- Vanadium
- Zinc

Other Associated Materials:

- Benzene


## Wetting Agents

Raw Materials, Intermediate Products, Final Products, and Waste Products
Generated During Manufacture and Use:

- Benzyl chloride
- Ethylenediamine


## White Metal

Raw Materials, Intermediate Products, Final Products, and Waste Products
Generated During Manufacture and Use:

- Antimony
- Tin


## Wicks

Raw Materials, Intermediate Products, Final Products, and Waste Products
Generated During Manufacture and Use:

- Boron


## Window Cleaning Fluid

Raw Materials, Intermediate Products, Final Products, and Waste Products
Generated During Manufacture and Use:

- Isopropyl alcohol
- Propyl alcohol

### Wine Making

Raw Materials, Intermediate Products, Final Products, and Waste Products Generated During Manufacture and Use:

- Formic acid
- Sulfur dioxide

### Winther's Acid

Raw Materials, Intermediate Products, Final Products, and Waste Products Generated During Manufacture and Use:

- alpha-Naphthylamine

### Wire Coatings

Raw Materials, Intermediate Products, Final Products, and Waste Products Generated During Manufacture and Use:

- Isocyanates
- Tin

### Wire Drawing

General Types of Associated Materials:

- Alkalis
- Oils
- Soaps

Raw Materials, Intermediate Products, Final Products, and Waste Products Generated During Manufacture and Use:

- Sulfuric acid

## Wood Bleaches

Raw Materials, Intermediate Products, Final Products, and Waste Products Generated During Manufacture and Use:

- Oxalic acid
- Sulfur dioxide

## Wood Burning

Raw Materials, Intermediate Products, Final Products, and Waste Products Generated During Manufacture and Use:

- Creosote

## Wood Preservatives

General Types of Associated Materials:

- Gasoline
- Resins
- Tar and derivatives

Raw Materials, Intermediate Products, Final Products, and Waste Products Generated During Manufacture and Use:

- Arsenic
- Benzidine
- Boron
- Chlorinated naphthalenes
- Chlorodiphenyls
- Chromium
- Copper
- Creosote
- Cresol
- Dinitrophenol
- Fluorides
- Formaldehyde

Raw Materials, Intermediate Products, Final Products, and Waste Products Generated During Manufacture and Use (cont.):

- Mercury
- Mercury, alkyl
- Pentachlorophenol
- Phenols
- Phenols, chlorinated
- Selenium
- Sulfur chloride
- Tetramethylthiuram disulfide
- Zinc

Other Associated Materials:

- Toluene
- Xylenes

## Wood Working

General Types of Associated Materials:

- Acids
- Alkalis
- Bleaches
- Epoxy resins
- Fillers
- Glues
- Lacquers
- Paints
- Petroleum hydrocarbons
- Solvents
- Stains
- Varnishes
- Waste oils

Raw Materials, Intermediate Products, Final Products, and Waste Products
Generated During Manufacture and Use:

- Chlorinated lime
- Formaldehyde
- Heavy metals
- Phenols
- Rosin

Other Associated Materials:

- Furfural
- Titanium

## Wool Processing

General Types of Associated Materials:

- Calcium chloride
- Naphtha

Raw Materials, Intermediate Products, Final Products, and Waste Products
Generated During Manufacture and Use:

- Bacteria
- Cresol
- Dibromomethane
- Formic acid
- Sulfuric acid
- Tetrachloroethene

## X-ray Tubes

Raw Materials, Intermediate Products, Final Products, and Waste Products Generated During Manufacture and Use:

- Mercury
- Molybdenum
- Titanium

## Xanthogenates

Raw Materials, Intermediate Products, Final Products, and Waste Products Generated During Manufacture and Use:

- Carbon disulfide

## Xerography

Raw Materials, Intermediate Products, Final Products, and Waste Products Generated During Manufacture and Use:

- Dichloroethane

## Xylenes

General Types of Associated Materials:

- Naphtha

Raw Materials, Intermediate Products, Final Products, and Waste Products Generated During Manufacture and Use:

- Phenols
- Pyridine
- Thiophene
- Trimethyl benzene

## Xylol

• See Xylenes

## Yeast

Raw Materials, Intermediate Products, Final Products, and Waste Products Generated During Manufacture and Use:

• Acetaldehyde

## Zectran

General Types of Associated Materials:

• Carbamate insecticides

Raw Materials, Intermediate Products, Final Products, and Waste Products Generated During Manufacture and Use:

• See Insecticides

## Zineb

General Types of Associated Materials:

• Dithiocarbamate fungicides

Raw Materials, Intermediate Products, Final Products, and Waste Products Generated During Manufacture and Use:

• See Fungicides

## Zinc Processing

Raw Materials, Intermediate Products, Final Products, and Waste Products Generated During Manufacture and Use:

- Cadmium
- Lead
- Platinum
- Potassium hydroxide

## Ziram

General Types of Associated Materials:

- Dithiocarbamate fungicides

Raw Materials, Intermediate Products, Final Products, and Waste Products Generated During Manufacture and Use:

- See Fungicides

## REFERENCES

Cline, P., and Viste, D., 1984, Migration and Degradation of Volatile Organic Compounds, in Management of Uncontrolled Hazardous Waste Sites: Hazardous Materials Control Research Institute.

Domenico, P., and Schwartz, F., 1990, Physical and Chemical Hydrogeology: John Wiley & Sons.

Environmental Protection Agency, 1980, Onsite Wastewater Treatment and Disposal Systems.

Ford, K., and Gurba, P., 1984, Methods for Determining Relative Contaminant Mobilities and Migration Pathways Using Physical-Chemical Data, in Management of Uncontrolled Hazardous Waste Sites: Hazardous Materials Control Research Institute.

Funderburk, J., III, 1990, Site Assessments Call For Variety of Approaches: Hazmat World, Vol. 3-90.

Hawley, G., 1987, Condensed Chemical Dictionary, Van Nostrand Reinhold Company.

Howard, P., 1990, Handbook of Environmental Fate and Exposure Data for Organic Chemicals, Vol. 2, Solvents: Lewis Publishers, Inc.

_____, 1991, Handbook of Environmental Degradation Rates:   Lewis Publishers, Inc.

Kenaga, E., and Goring, C., 1980, Relationship Between Water Solubility, Soil Sorption, Octanol-Water Partitioning, and Concentrations of Chemicals in Biota, in Aquatic Toxicology, ASTM STP 707, J. G. Eaton, P. R. Parrish, and A. C. Hendricks, Eds., American Society for Testing and Materials.

Key, M., Henschel, A., Butler, J., Ligo, R., Tabershaw, I., and Ede, J., eds., 1977, Occupational Diseases - A Guide to Their Recognition: U.S. Department of Health, Education, and Welfare.

Kramer, J. and Herbert A., 1988, Metal Speciation - Theory, Analysis, and Application: Lewis Publishers, Inc.

Mauch, J., 1990, Site Assessment Standards Sorely Needed: Hazmat World, Vol. 3-90.

McGregor, G., 1991, Being "Too Helpful" Makes Consultants Liable for Contaminated Real Estate: Hazmat World, Vol. 3-90.

Michigan Department of Natural Resources, 1989, Michigan Sites of Environmental Contamination Proposed Priority Lists, Act 307.

Montgomery, J., and Welkom, L., 1989, Ground Water Chemicals Desk Reference: Lewis Publishers, Inc.

Montgomery, J., 1991, Ground Water Chemicals Desk Reference, Volume 2: Lewis Publishers, Inc.

Neufeldt, V., and Guralink, D., eds., 1988, Webster's New World Dictionary, Webster's New World.

Prescott, M. and Brossman, D., 1990, The Environmental Handbook for Property Transfer and Financing: Lewis Publishers, Inc.

Salveson, R., 1984, Downtown Carcinogens - A Gaslight Legacy, in Management of Uncontrolled Hazardous Waste Sites: Hazardous Materials Control Research Institute.

Schaumberg, F., 1990, Banning Trichloroethylene: Responsible Reaction or Overkill: Environmental Science Technology, Vol. 24, No. 1.

Sittig, M., 1981, Handbook of Toxic and Hazardous Chemicals: Noyes Publications.

Small, M., 1977, Natural Sewage Recycling Systems: National Technical Information Service.

Smith, L., and Dragun, J., 1984, Degradation of Volatile Chlorinated Aliphatic Priority Pollutants in Groundwater: Environment International, Vol. 10, pp. 291-284.

Windholz, M., ed., 1989, The Merck Index, 11th edition: Merck & Company.

APPENDIX A

Transformation Products of Common Contaminants

The occurrence of exotic and uncommon chemicals in the environment can be perplexing to environmental professionals, landowners, and purchasers of property when no obvious uses or sources of these chemicals can be identified.    An understanding of transformation products of common contaminants is especially important for those involved in ground water quality monitoring.    Recent chemical and environmental research has demonstrated that contaminants may undergo significant transformations in the environment.    Significant ongoing research is being conducted in the area of bioremediation and will undoubtedly enhance our knowledge of this subject.    Transformation mechanisms will, therefore, be discussed only briefly.    The data presented represents a compilation of known and suspected transformation products of common contaminants.

The information presented in this appendix will be limited to documented transformations of common contaminants in soils and ground water.    These data are generally regarded as correct, however, site-specific conditions must be evaluated to determine if conditions are or were favorable for chemical transformations to have occurred.    The reader is referred to Howard, et al. (1991) for the degradation rates of these and other compounds in various media.

## Chlorinated Compounds

Chlorinated hydrocarbons are environmentally mobile, persistent compounds widely used as solvents and in various industrial processes.    The mobility and persistence of these compounds often results in the discovery of environmental contamination from releases or disposal practices that may have occurred many years previously.    Transformation products of these solvents are also frequently found at sites of environmental contamination.

It is generally accepted that chlorinated hydrocarbon compounds undergo dehalogenation under various anaerobic conditions (Cline and Viste, 1984) by microbial and chemical processes (Smith and Dragun, 1984). Dehalogenation essentially involves the liberation of a $Cl^-$ ion from the parent compound.    The dehalogenation phenomena is important as it can be used to reasonably explain how contamination of a resource by these products has occurred when there was no obvious explanation for their presence.    The presence of these compounds may also be used to infer the historical use of a more common compound. Recent, as yet undocumented, studies suggest that these chemicals may be further broken down microbially under aerobic conditions.

### Chlorinated Benzenes

Chlorinated benzenes may enter the environment through their common use as solvents and industrial chemicals. O−Dichlorobenzene transforms to chlorobenzene upon the loss of a $Cl^-$ ion. Chlorobenzene may transform to 2−,4−chlorophenol in soil and water.

### Chlorinated Ethanes

Chlorinated ethanes are likely to enter the environment through air emissions, waste water discharges, and spills through their use as solvents, degreasers, and chemical intermediates. 1,1,1−Trichloroethane may transform to 1,1−dichloroethane, 1,1−dichloroethene, or trans 1,2−dichloroethene upon the loss of a $Cl^-$ ion. 1,1−dichloroethane transforms to chloroethane upon the loss of another $Cl^-$ ion. 1,1−Dichloroethene transforms to vinyl chloride with the loss of another $Cl^-$ ion.

### Chlorinated Ethenes

Tetrachloroethene, also known as perchloroethene, and trichloroethene are commonly used as solvents and degreasers in a number of industries including dry cleaning, plating operations, and manufacturing. Tetrachloroethene transforms to trichloroethene upon the loss of a $Cl^-$ ion. Trichloroethene transforms to cis 1,2− and trans 1,2−dichloroethene and then to vinyl chloride upon the loss of subsequent $Cl^-$ ions.

### Chlorinated Methanes

The compounds of the chlorinated methane transformation series are all commonly used industrial contaminants. They are widely used as solvents and have many uses in various industries. Tetrachloromethane (carbon tetrachloride) transforms to chloroform and then to methylene chloride upon the subsequent loss of $Cl^-$ ions.

## Aromatic Hydrocarbon Compounds

Aromatic hydrocarbons commonly enter the environment through their use in petroleum fuels, such as gasoline, and through their use as solvents. Common aromatic hydrocarbons for which transformation compounds have been suggested include benzene, ethyl benzene, and toluene. Possible transformation products of benzene include phenols and nitrophenols, dihydroxybenzene, nitrobenzene, formic acid, and catechol. Ethyl benzene may transform to such compounds as acetophenone and benzaldehyde. Toluene may transform to compounds including acetic acid, benzaldehyde, benzyl alcohol, catechol, and cresol.

## Other Common Contaminants

Acrylonitrile is commonly used in plastics, rubbers, and organic synthesis. Acrylonitrile may be transformed to formaldehyde.

Nitrobenzene is used in various manufacturing processes and solvents. Nitrobenzene may also be a transformation product of benzene. Nitrobenzene may be transformed to aniline and benzidine.

Phenols are produced from coal tar and are used in the production of resins and explosives. Phenols may be transformed to acetic acid, catechol, quinones and hydroquinones, and tri-, di-, and chlorophenols.

## APPENDIX B

### Determining the Mobility of
### Potential Environmental Contaminants

**Organic Compounds**

The migration potential of organic contaminants can be assessed using a calculated mobility index. The mobility index (MI) is a relative measure of a contaminant's relative tendency to migrate in the environment and reflects a contaminant's migration potential in water, air, and soil (Ford and Gurba, 1984). The mobility index can be represented as:

$$MI = \log \frac{Water\ Solubility \times Vapor\ Pressure}{Soil\ Sorption\ Coefficient\ (Koc)}$$

where Soil Sorption Coefficient (Koc) is the concentration of the chemical sorbed by the soil expressed on a soil organic carbon basis divided by the concentration of the chemical in the soil water (Kenaga and Goring, 1980).

Readers are referred to both Kenaga and Goring (1980) and Ford and Gurba (1984) for detailed discussions regarding the determination of Koc and for tables of calculated values for various compounds. Montgomery (1991) also contains Koc data for a substantial number of compounds. It is important to note that the solubility of organic compounds can be dependent upon temperature.

The following table is a general guide to the Mobility Index and the relative mobility of organic compounds.

| Relative Mobility Index | Mobility Descriptions |
|---|---|
| >5.00 | Extremely Mobile |
| 0.00 to 5.00 | Very Mobile |
| -5.00 to 0.00 | Slightly Mobile |
| -10.00 to -5.00 | Immobile |
| <-10.00 | Very Immobile |

## Inorganic Compounds

The mobility, or conversely the persistence, of inorganic compounds is a function of a number of variables. The majority of inorganic compounds are not likely to move from locations in which they were formed or deposited unless acted upon by ground water or other fluids. The degree to which inorganics will dissolve, migrate, and redeposit is dependent upon a numbers of variables such as eh, pH, temperature, and the availability of ionic catalysts. Complexation and other types of reactions may facilitate the transport of various compounds. Analysis of site-specific conditions are necessary to determine the mobility or persistence of inorganic environmental contaminants. Readers are referred to Kramer and Allen (1988) for in-depth information regarding the fate of inorganic compounds in the environment.

## APPENDIX C

### Additional Information Sources

American Chemical Society
1155 16th Street, NW
Washington, DC  20036

Chemical Abstracts Service
2540 Olentangy River Road
P.O. Box 3012
Columbus, OH  43210

Chemical Information System, Inc. (CIS)
Computer Sciences Corporation
6565 Arlington Blvd.
Falls Church, VA  22046

ChemShare Corporation
P.O. Box 1885
Houston, TX  77001

DIALOG Information Service, Inc.
3460 Hillview Avenue
Santa Barbara, CA  93103

National Institute for Occupational Safety and Health
4676 Columbia Parkway
Cincinnati, OH  45226

OSHA Analytical Laboratory
P.O. Box 15200
1781 S. 300 West
Salt Lake City, UT  84115

SDC Information Services
2500 Colorado Blvd.
Santa Monica, CA  90406

Technical Database Services, Inc.
10 Columbus Circle
Suite 2300
New York, NY 10019

U.S. Dept. of Health and Human Services
4676 Columbia Parkway
Cincinnati, OH 45226

U.S. Environmental Protection Agency
Chemical Information Branch
401 M Street, S.W.
Washington, DC 20460

9 780367 450366